The Other Side of the Medal

The Other Side of the Medal

A Paleobiologist Reflects on the Art and Serendipity of Science

Everett C. Olson
*Professor of Zoology Emeritus
University of California
Los Angeles, California*

The McDonald & Woodward Publishing Company
Blacksburg, Virginia
1990

The McDonald & Woodward Publishing Company
P. O. Box 10308, Blacksburg, Virginia 24062-0308

The Other Side of the Medal

©1990 by The McDonald & Woodward Publishing Company

All rights reserved
Printed in the United States of America
by BookCrafters, Inc., Chelsea, MI
Composition by Marathon Typesetting, Roanoke, VA

10 9 8 7 6 5 4 3 2 1

First printing April 1990

Library of Congress Cataloging-in-Publication Data

Olson, Everett Claire, 1910–
    The other side of the medal : a paleobiologist reflects on the art and serendipity of science / by Everett C. Olson.

    p. cm.
    ISBN 0-939923-13-0 : $22.95
    1. Olson, Everett Claire, 1910–   . 2. Paleontologists—United States—Biography. 3. Paleontology—Permian. I. Title.
QE707.044A3   1990
560.9—dc20

[B]   90–5612
                                                                               CIP

Reproduction or translation of any part of this work, except for short excerpts used in reviews, without the written permission of the copyright owner is unlawful. Requests for permission to reproduce parts of this work, or for additional information, should be addressed to the publisher.

Everett C. Olson
(ca. 1986)

Dedication

To all of those graduate students who worked with me during the past 50 years; whose brilliance, challenges, knowledge and irreverence guided my intellectual growth and kept me young in mind; whose strong arms, willing backs and walking legs made possible our explorations, collecting and excavating of fossil vertebrates when found; who have gone forward to productive careers in their own fields and of their own choices.

Table of Contents

	Preface	ix
1	The Permian and I	1
2	The Arroyo Formation, Coffee Creek and Ernest	17
3	Eyes West—to Younger Beds	27
4	Ignorant Ridge—the Vale and Choza Formations	37
5	Land, Weather and Dogs	51
6	The San Angelo Formation—and a Look Backwards	63
7	Eyes to the East	79
8	Impressions—Efremov, Museum, Friends	97
9	Efremov—the Man	111
10	Efremov, Science and Philosophy	119
11	Frivolity, Dialectical Materialism and Science	129
12	The Other Side of the Medal	147
13	Books, Writings and Ideals	161
14	What of Dreams—Now?	173
	Index	181

Preface

The narrative of this book is set in the years from 1935 to 1972. It is written in an autobiographical mode to record some of the events that attended my efforts to learn what I could about what life on land was like during a remote time known as the Permian period. My studies led along two pathways, widely disparate in time but parallel and intimately associated. One followed the evolution of life through the 50 million years recorded in the rocks and fossils of the Permian, which began about 280 million years ago. The other covered the few decades during which people, places and mores, and ideas molded my views of life. The story of these pathways is simply one that I want to tell.

My account begins with what I have called "midpoint," the breakpoint between my work in the field and laboratories of the United States and my ventures in the museums of the Soviet Union. Texas is the setting of the first part of the narrative, with two "cowboys"—Ernest Cruthirds and Wade Barker—as central characters. There in Texas, I and my parties lived the life typical of fossil hunters of the mid-decades of the 20th century. We learned a lot about the ancient life of the Permian and its evolution. After the "midpoint," although the Texas work continued, the setting of the narrative is mainly the Soviet Union and the central character is Professor Ivan A. Efremov (pronounced Ye-FRE-mov). During this period, from 1959 to 1972, while fossils continued to play an important role in my life, more and more a shift to a new intellectual level emerged. The first part of the narrative, therefore, concerns events that were fairly simple and uncomplicated, whereas the second in places deals with more complex and elusive questions and thoughts, most of which are without resolution.

Basically, the story I want to tell ends with the death of Efremov in 1972. Although he is the central figure in the later chapters, probably few who read this will ever have heard of him and, even if they have, may know little about him. So why do I write about Efremov in what is otherwise a personal narrative? Partly because he has been important to me as a friend and scientist, and partly because he has been instrumental in expanding my intellectual horizons. But also, he should be much better known beyond the Soviet Union both as a writer and a scientist. I hope that what is included in this book will help to remedy this lack of knowledge of his works.

Professor Ivan A. Efremov was an outstanding scientist, one of the great paleontologists of the Soviet Union. Although he wanted to be known as a scientist, he would modestly cover his accomplishments with denials except when extolling them in off moments. Only since about 1960 has his work begun to be appreciated in the Western world, and I hope that my early efforts in making his studies more accessible have had some influence in this direction.

Professor Efremov's scientific studies were devoted primarily to the geology and the animal life of a time in earth history known as the Permo-Triassic, which covered some 60 million years between about 260 and 200 million years ago. Almost singlehandedly he revolutionized the studies of this time period in the Soviet Union.

During the Permo-Triassic, the ancient amphibians, which were very different from the frogs and salamanders of today, began their decline, after having reached great prominence. Reptiles, on the other hand, were casting off the old and bringing on the new. This transformation of reptiles culminated in two major lines, one leading to our own group, the mammals, and the other to snakes, lizards and birds of today as well as to the now extinct dinosaurs and flying reptiles. No satisfactory time scale had been developed for the fauna of the Permo-Triassic of the Soviet Union until Efremov provided one. Classifications of many of the amphibians and reptiles were chaotic. Efremov remedied this, though, of course, not to the satisfaction of everyone. He made detailed studies of the deposition and preservation of fossils under the heading of taphonomy, a term which he introduced. His studies were recorded in some 100 essays, treatises and books among which the following are particularly important:

The Fauna of the Terrestrial Vertebrates in the Permian Copper Sandstone of the Western Cisuralian Region. Trudy Paleontological Institute, volume 54, 416 pp. Academy of Sciences USSR, 1954. In Russian.

Catalogue of the Localities of Permian and Triassic Terrestrial Vertebrates in the Territories of the USSR (with B. P. Vjuschkov). Trudy Paleontological Institute, volume 46, 147 pp. Academy of Sciences USSR, 1955. In Russian. (English Summary: E.C. Olson, *Journal of Geology*, volume 65, pp. 196–226, 1975.)

Taphonomy and the Fossil Record. Trudy Paleontological Institute, volume 24, 178 pp. Academy of Sciences USSR, 1950. In Russian. (French translation: S. Ketchian and J. Roger, *Annalles du Centre d'Etude et de Documentes Paleontologie*, no. 4, 164 pp., 1953.)

The Trail of the Winds. All Soviet Scholarly-Pedagogical Publishing House, Moscow, 1956, 366 pp.

The earliest part of the time represented in Efremov's principal studies overlapped that of the last part of mine, and it was this common interest in the paleontology of the Upper Permian that first brought us together. As our scientific contacts enlarged and a personal association emerged, another facet of Professor Efremov came to my attention. Beyond his science, he was a dreamer, and he found an outlet for this facet of his personality in fictional writing—science fiction, fantasy, romance and novels. Ironically, Efremov is better known outside of the Soviet Union for his science fiction writings than for his science. Many of his stories and books have been translated—one, "Andromeda," into 35 languages. In English, most of these stories give a somewhat false image, for some of the translations are less than adequate, and two works which I consider among his best—*The Razor's Edge* and *The Hour of the Bull*—are available only in Russian. Many of Efremov's social and philosophical concepts appear in these works, couched in a Russian that is difficult reading for a non-Russian, with its bows to the "older" language which he revered and use of words that are found only in the most inclusive dictionaries.

This, then, is the man who came to mean so much to me as our friendship developed and whose ideas and musing, both scientific and philosophical, formed a mirror that reflected the dialectic of our development in the 1960s as we came from disparate cultures and circumstances to reach a common understanding in both social and philosophical realms. How this came about is a main theme of later parts of this book.

The pathways to our coming together were diverse and, for both of us, somewhat bumpy. By the time I had attained my Ph.D. and begun to teach geology and paleontology at the University of Chicago, in 1935, the world was engulfed in a major depression. During the ensuing three and a half decades covered in this book, the world remained in turmoil, with short periods of seeming stability interspersed freely with wars, drought, famine and immense restlessness and danger. Relationships between the United States and the Soviet Union, rarely untroubled, went from cool to cold, ameliorated only by short warming spells. Life in academia, where I was, had a degree of stability that was seriously disrupted only during World War II and in the McCarthy era of the early and middle 1950s.

During most of the years from about 1935 to 1965, and those earlier in which I was a university student in geology and paleontology, the intellectual world of earth scientists was cradled in a comfortable framework of uniformity, fixed continents and the Neodarwinian theory of evolution. This last paradigm—formed by a fusion of Darwin's concept of natural selection and 20th century genetics—provided the structure within which most paleontologists interpreted the fossils preserved in the rocks deposited during some three billion years of the four and a half billion years of the existence of the earth.

Cracks in the structure of the intellectual framework of earth scientists had begun to appear in the late 1950s at about the time that I first began to recognize a close relationship between the animals of the Permian of North America and those of Russia. This relationship became a crucial focus of my later studies. A developing, dynamic point of view about the earth and its continents presaged the oncoming revolution. It dispelled the notion that the "self evident" processes of slow contemporary change, as then known, could provide a framework that could be cast back to give explanations of ancient life. The intuitive human time frame was inappropriate.

In contrast to the established point of view, the continents came to be seen as restless, drifting islands carried on rigid plates that made up the outer part of the earth. Disturbingly new ways of looking at the fossil record began to appear. Even by the late 1960s, however, the impacts of such "unorthodox" ideas were just starting to be felt by the majority of paleontologists.

In a sense then, this story starts in an "age of innocence" and ends as the protecting cover of a comfortable paradigm is rudely peeled away. But the eventual effects of any such changes are hard to evaluate as they are taking place. Drastic as they seem today, they may appear quite differently tomorrow. Also, as the vast reorganization of geological thinking was taking place, the complacency about the sufficiency of the Neodarwinian theory of evolution, or the Synthetic theory in its broad guise, was being questioned, even as its position hardened. This theory in its simplest terms combines Darwin's natural selection and the data of modern genetics into a coherent explanation of evolution. Over the last two decades some scientists have been probing the outer shell of this paradigm. Will this basis for interpretation also be shattered?

It was matters such as these that Professor Efremov and I pondered from our diverse backgrounds. Over both of us the "old guard" had held tight sway and taught those of us who entered geology during the "age of innocence" to follow their ways. I and my generation of the "middle guard" of the time wavered, but with some reluctance, and formed, we would like to think, a balance wheel to even younger scientists who, as we before, knew little of the past and began with the bright confidence that what they had been taught about the new interpretations was the final truth. So at about the time this narrative ends, an exciting new stage was beginning under circumstances far different from those that existed in 1935, but the "new guard" entered "its" stage with no less confidence than that felt by those of us who began about then only to be rudely wrenched from our complacencies.

At the very end of my story of travel through time I speculate about what the central figure of the last part of the story, Professor Efremov, might have thought had he lived until now. Would he have been a dour rejector of the new or would he have adjusted and set his sights anew, starting from his unique base of dialectical thought?

The Other Side of the Medal

1

The Permian and I

Midpoint

Some 20 long hours out of Chicago and New York, I was dozing to the drone of the motors of the KLM DC 6. The Captain broke in, "We will be landing in Moscow in 25 minutes." "What on earth am I doing here?" flashed through my half-awake mind. Ahead was the formidable Russian city of Moscow, behind was a comfortable house, my wife and children worrying about me, a good profession and colleagues.

This was May 16th, 1959. The Soviet Union had a hostile image. But coming to Moscow had seemed to me the way to cap my work on ancient reptiles and amphibians, by comparing the ancient Permian fossils of North America and those of the western flanks of the Ural Mountains of Europe. The Paleontological Museum in Moscow was the magnet, but now that the city and museum were looming ahead and I was travel-weary, I approached them with some trepidation.

We did land, but not before the pilot had said, "Well, there it is, folks, that's where you will stay," as if somehow he was glad he would not be there long. The city from 5000 feet loomed a misty brown, outlined by dim lights. Once on the ground, we stopped in a light rain some distance from the reception area. My companions in flight, a group of 13 British automation experts, deplaned and were met with hugs and kisses from the Russians. I was met at the door of the plane by a military man who took my passport and said brusquely, in Russian,

"Let's go." The British 13 formed a chattering procession with their friends, and I, with my militia man, a sober tandem group of two.

At the dimly lit reception desk the clerk took my credentials and was looking them over when I said, in what I hoped was Russian, "I am Professor Olson, from Chicago." Over my shoulder came a voice in English, "And I am Professor Orlov, from Moscow." The nicest words I have ever heard! He said something to the clerk in Russian, and to me, "Let's have a beer." We did, and then departed for the Hotel Ukraine (Gastinitza Ukraina) in the Paleontological Institute's car. Professor Orlov was the head of the museum and its only Academician at that time. He took me to the 30th floor of the hotel and, after bacon and eggs at 2:30 A.M., I found my room and crawled in between the heavy Russian "comforters."

Thus began one of the most pleasant and productive times I have ever had while visiting a museum for study purposes. I didn't know then that this was to be but the first of seven visits or that I would find the wealth of Permian fossils which so neatly complemented those we had been finding in the middle part of the Permian of North America. The foundations for the study had been laid in research which started in 1935 in the earlier Permian beds of north central Texas. Moscow proved to be a major turning point in my investigations of the fossils as well as how I looked at the worlds of the both Permian and today.

Beginnings

The turbulent Permian period of geological time ended in massive extinctions of life on the land and in the seas. Long ago, beginning some 280 million years ago and lasting for about 50 million years, a vast single ocean and one massive continent set the stage for the final act of the evolutionary drama of the Paleozoic era. Some animals and plants did survive the early holocaust and some became our ancient ancestors. How this all came about, what it means, and forays into the special realms of historical science, with the Permian at center stage, have been the focus of my lifelong researches. How I threaded my way along the twistings and turnings of a search for knowledge and the friends I met along the way are the subject of this narrative.

When I entered college in 1928, I had no idea that one day I would become a student of "old bones"—a vertebrate paleontologist. I didn't even know what one of that ilk was, for unlike the youngsters of today, we were not "dinosaur nuts." Nor did I dream that a pair of cowboys, Texas villagers and ranchers, and a Russian writer and dreamer, Professor Ivan A. Efremov, would guide my pathways as I left my mentor, Professor Alfred S. Romer, with a shiny new Ph.D. in 1935. Were it not for "old bones," the stimulation of the University of Chicago, and many friends and helpers my life would have turned out differently. The Permian just happened to be the time in which I landed.

I first saw the light of day on November 6, 1910. A few years later I, like most children, became a bundle of curiosity and questions. Beyond pestering my parents with questions about the nature of the puzzling matter of time, my earliest "want to know" that I can recall led me to collect and crudely classify butterflies. It was in 1914 or 1915, in Berlin, Wisconsin, where my mother was tending her mother, who was ill, that I became fascinated by the big brown butterflies (monarchs) that perched on sunny days on red and yellow zinnias between my grandparents' house and the home of Dr. Walbridge to the south. Along with the stately monarch, which could be picked off by hand, I managed to populate the screened porch with woolly bear caterpillars who wandered across the floor, sometimes getting squashed, and climbed the screens. The older people were tolerant.

For the next several years, until time for college, I kept at this hobby, and today, once again, my wife and I wander the world as amateur lepidopterists. With a close friend of my youth, Dudley Knox, I hunted the summer fields, collecting moths and butterflies, caterpillars for my menagerie, beetles, plants and anything else that piqued my curiosity about nature. In 1927, a trip to Florida with my co-collector Dudley capped this phase. I was 16 and he 15 as we took off in a 1921 Model T Ford, making about 150 miles a day. How our parents let us go remains a mystery to me.

My family lived then in Hinsdale, a small suburb of Chicago, in a big house with a big yard and barn. My father, Claire Myron Olson, was a dentist with a practice in Chicago and Hinsdale, and my mother, Aimee Hicks Olson, gave me and my older brother, Guindon Olson, a fine, comfortable and stable environment in which to grow up. We were very lucky in our parents. The years in Hinsdale were kindergarten, grade school and high school; piano, violin and piccolo lessons; athletics, first in the backyard and then high school football, basketball and track. At 5 feet 6 inches I was far from a power in contact sports but good enough to earn some letters. Grades were never much of a problem, mainly, as I recall, because I was scared of the teachers. I turned out to be valedictorian of my class in high school when the girl who should have been so honored ran off and got married just before graduation. This was taboo under the mores of the times.

We had some good and some poor teachers at Hinsdale Township High. One of the best, Mr. Brooks, introduced me to chemistry and for a while I thought I had found my academic niche. But another teacher, in mathematics, was not for me and he, along with my adviser, failed to tell me that my mathematics was insufficient not only for chemistry but for any branch of college science. I was really much better in Latin, and so took the University of Chicago scholarship examination in this field. The Virgil of the small town boy was no match for the Ovid of the intensely tutored students from the Chicago schools. But I was instead awarded a two-year athletic scholarship, an occasion for amusement later among my colleagues, many jaundiced with respect to college athletics. It so happened that gymnastics became my forte. I finally became quite good at bars, rings and mats. For two years I captained the team, and we won the Big Ten championships three years in a row. The University of Chicago was still a power in sports in the late 1920s and early 1930s, when athletics began to take a back seat to scholastic achievements. I amassed a nice batch of medals, the best being the Big Ten Athletic-Scholastic medal in my senior year. Phi Beta Kappa, into which I slipped by the barest margin, was the other side of the medal.

Disenchanted with chemistry after six or seven "memorization" courses, I switched to geology, drawn by the inspiration of Professor J Harlan Bretz. He was a hard headed Germanic professor who taught not by giving out information but by demanding reading and by challenging students in class, often bitterly and sarcastically. I loved to spar with him in class, rarely winning but always leaving with a desire to know more.

The later part of my undergraduate career, 1930–1932, merged with the depression of the 1930s. Things continued to worsen; as we entered graduate school on a shoestring, any thoughts my fellow students or I may have had earlier about the commercial value of a degree vanished. In a way this was good. Had I been thinking along such lines I never would have ended up in vertebrate paleontology. This esoteric study of extinct animals has rarely promised certain employment, let alone the possibility of wealth. It was education for education's sake, which is not all bad.

After receiving an S.B. degree in general geology and paleontology, I continued my education with work toward a master's degree in invertebrate paleontology under Professor Carey Croneis. This was in 1933 and things were tough financially. Many of us from the Department of Geology at the University of Chicago had found jobs at the Century of Progress—the 1933 Chicago World's Fair—aided by Professor Croneis, who was in charge of the Hall of Science. I was the partner of Llewelyn Price, another paleontology student, in an exhibit on evolution, which was manned seven hours a day, seven days a week, at $25 per week—a godsend.

Seeing no place to go during this time, I wrote some 50 letters to small colleges, telling them I could teach geology and "gym," then a compulsory subject. I received three replies and one solid offer!

Just as I was going to accept the offer, the Department of Geology gave me one of the four departmental fellowships. It covered tuition and left $250 to be spent anyway one wished. Unlike most fellowships today, this one required work. My duties in 1934 and 1935 consisted of reviewing books, and one of the odd results of this duty is that my most extensive annual list of publications is for 1935. The entries were neither the best nor the most significant, but they were plentiful. The editor of the *Journal of Geology*, which published the reviews, had great faith in me for he sent me books in French, German, Spanish and Italian, as well as English. As a result, I lost my fear of foreign languages as far as the written word was concerned.

The tedium of the summer's work at the World's Fair in 1933 was such that I felt I just couldn't stand another season of it. Lawn work at 25 cents per hour (35 cents for really hard jobs) and tutoring "any high school subject" at $1.00 per hour brought in enough cash to allow me to continue my education into 1935. By then I had switched over into vertebrate paleontology.

At the University of Chicago, in 1933, Professor Alfred S. Romer had initiated me into the marvels of Permian reptiles and amphibians. Romer was then following in the footsteps of such illustrious predecessors as professors Ermine C. Case and Samuel W. Williston, who studied the Permian before him. The wooden case drawers of the University of Chicago's Walker Museum were filled with unique collections of animals from the red beds of Texas, accumulated in large part by Paul C. Miller, Curator of Vertebrates in Walker Museum. With Romer and Miller and the outstanding collections, it was a heady introduction to the vertebrate paleontology of the Permian for a youngster of only 23.

What happened was this. Professor Romer—Al, and a close friend in later years—and I had come together in one of those odd ways that change careers. I had taken his course in Vertebrate Paleontology as a junior in college, and then strayed over into invertebrates. In 1933, when prohibition was on the way out, the Geological Society of America met at the University of Chicago. A big reception and smoker was held under Mitchell Towers. There were two punchbowls, one spiked with ersatz gin, the other pure. Good paleontologists never passed up a bit of stimulation, so by early evening both Romer and I found our inhibitions of student and faculty member relaxed. With my arm around his shoulder I slurred out that I thought I would take a research course with him. He slurred back, "Thash good." I did, and the switch was made.

I had just begun my studies of vertebrate animals under Professor Romer when, in 1934, he answered a call to Harvard and left me geographically alone, but at least under his remote guidance. About a year after this, the University of Chicago, with as much courage as good sense, but under urging from Romer, offered me the position of Instructor, to "fill" the shoes of my former mentor. I had just about completed my Ph.D. and was beginning to search for a job in those bleak times. Postdoctoral positions had not yet attained any prominence. The Department of Geology hurried me through my final degree requirements and Professor Romer did a massive job of revising my thesis. Then the University sent me on a six months tour of museums, universities and field localities, paying me a salary so I could afford to travel in my own Model A Ford. Once cast in this role, I was hooked, and my future was wedded to studies of the extinct animals of the Permian. Of course, it wasn't all straightforward.

When Professor Romer left for Harvard he borrowed a large suite of the Texas fossils in order to continue preparation of his monograph on Pelycosauria, primitive mammal-like reptiles. He left behind an important collection of later Permian fossils from the Karroo Desert of South Africa which included remains of advanced mammal-like reptiles, the Therapsida. Although my beginning field studies were in Texas, my early laboratory work was mainly devoted to the African materials that had been assembled some eight years before by Romer and Paul Miller. The happy result was that I developed an enduring interest in the evolution of the reptilian predecessors of the mammals and in the origin of our own ancestors. This interest motivated much of what I did thereafter.

The African materials, although from the Permian, were much younger than our Texas specimens and a major gap of several million years separated the evolutionary stages portrayed by these two collections. Collections in the Soviet Union partly, but not fully, filled this gap. As a result of language barriers and poor exchanges of information, we in the United States didn't know very much about the Russian specimens. Eventually this deficiency was remedied, but as I began my work on the African and Texas materials I had no idea that at some future time, all of the pieces would finally be brought together.

As time went by, and World War II was over, it became clear that if we ever were to relate the North American animals to those of the Soviet Union, and, through them, to the African materials, we must try to find fossils in America in beds that were higher and thus younger than any specimens so far known. With this need in mind, my parties and I pressed slowly into higher beds in Texas and then in Oklahoma, a venture in which we had considerable success.

This work in turn led to the development of my close scientific and personal relationship with Professor Ivan A. Efremov of Moscow, USSR. He, too, was interested in Permian life, and in our associations we found rapport in common ideas born in two highly diverse cultures. A substantial part of my narrative is devoted to our associations and their impacts on our common concerns about problems of science as well as society and philosophy, viewed from the perspectives of our mutual but often dissimilar interests in the intertwining lines of evolution and social development. But we both drew our major inspiration and sense of things from the rocks, times and animals of the 50 million years of Permian history. The reality and scope of such immense spans of time are the property of geologists, paleontologists and astronomers, and weave a peculiar bond in a world of thought incomprehensible to many in other fields.

The Permian

Permian times provide a common thread through all of this narrative and need some explaining. The Permian system and the place that it occupies in the calendar of geological history is shown in Figure 1. The relationships of the American, Russian and South African Permian include a minimum of detail necessary to make my remarks on the various subdivisions of time intelligible.

When I started exploring for Permian vertebrates in Texas, their fossils were known only from the lower beds, no higher than the Arroyo formation, the lowest rock unit of the Clear Fork group. These beds were first prospected in the late 1870s and extensive collections had been made and housed in many museums in the United States and abroad by the mid-1930s. But these activities took place long after the first remnants of vertebrates from the Permian were found in Russia. These date back to about 1780 and were unearthed by miners in their searches for copper. It was nearly 70 years later that the Permian was formally named. Some of the early specimens are still preserved, albeit in bad shape, in the Paleontological Museum in Moscow.

The Russian copper deposits occur in bands of red sandstone, shale and clay that stretch along the western face of the Ural Mountains from the Barent's Sea on the north to the Caspian Sea on the south (Figure 2). Centered in this belt, some 700 miles east of Moscow, is the city of Perm, the tangible namesake of the ancient Uralian Kingdom of Permia. During the 1840s Sir Roderick Murchison, a renowned British stratigrapher and paleontologist, had come to this region in the course of his studies of the geology of European Russia, commissioned by Nicholas the First of Russia. He named the red sedimentary sequence "Permian" and defined the system to include the beds that overlay the Carboniferous and underlay the Triassic (Figure 1). Gradually his Permian replaced other names for beds of this age and came to be accepted worldwide. The time during which the beds were deposited came to be known as the Permian period.

The Permian has turned out to be a fascinating time in earth history, one of massive changes in the animals and plants of the land and seas, during which the new great continent of Pangea approached its final shape. Biologically, the Permian formed the bridge between ancient life of the Paleozoic Era, which had reigned for almost 300 million years, and the new life of the following Mesozoic Era which, too, flowered, and then suffered a great extinction about 70 million years ago.

The Permian that I had studied in the 1930s and 1940s began to take on a very different look in the ensuing decades.

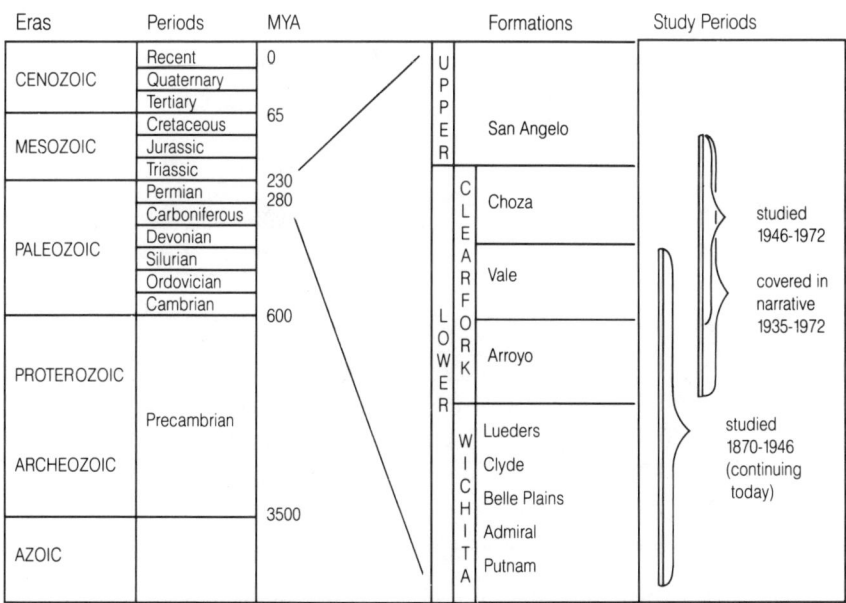

Figure 1. To the left, a general geological calendar, with approximate dates of boundaries indicated. To the right, the Permian section of part of Texas as covered in the text, with dates of studies of various formations up to 1972.

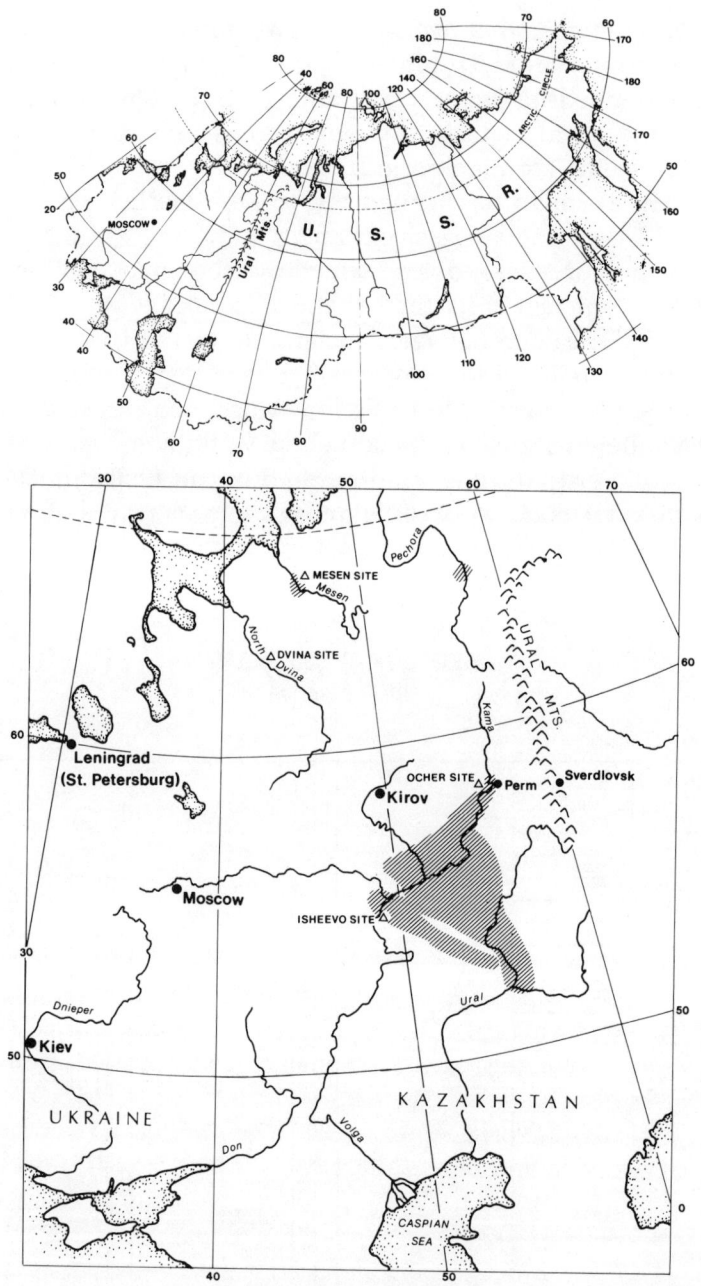

Figure 2. The USSR, with the western part of the country enlarged to show the location of places mentioned in the text. Permian outcrops are screened. (Enlargement modified from map in E. C. Olson, 1957, translation and condensation of I. A. Efremov, Catalogue of Localities of Permian and Triassic Terrestrial Vertebrates of the Territories of the USSR, *Journal of Geology* 65:196–226.)

For many years the idea that continents were fixed in their positions since the beginning of time had dominated geological and biological interpretations. True, there were some "heretics" who maintained that the continents had moved, but most of us rejected such "nonsense." Still, during the years before 1950, vague problems of animal and plant distributions and relationships nagged at some paleontologists. Then, as oceanographers and geophysicists began to explore the floors of the oceans and records of ancient magnetic fields preserved in the rocks, concepts of fixed continents became increasingly untenable. Clearly the sea floors had spread laterally from great mid-oceanic ridges and the records of change were preserved in the evidence of reversals of earth's magnetic field. The continents had been forced into movement and underwent massive changes in their alignments. The burgeoning data were formulated into a sweeping theory of plate tectonics, under which the crust of the earth is envisaged as being made up of a series of rigid plates which move slowly, in our perspective, but inexorably.

For many of us during the 1960s plate tectonics was a shattering concept. We were prone to reject it, for it shook the whole basis of geological theory. Western Europeans found it easier to accept than many others, although the Russians lagged! In North America, the east coast came along faster than the bastion of geological conservatism of the Midwest where I had been nurtured. It is terribly hard to give up an ingrained paradigm that has served so well.

Like many of my colleagues who studied ancient vertebrates of the Permian and earlier times, I had been a bit uneasy about fixed continents, for they did pose serious problems in intercontinental migrations of ancient cold-blooded animals. We knew that somehow animals and plants had passed rather freely between North America and Europe and also between South America and Africa. Land bridges did not satisfactorily solve the problem and ocean rafts seemed woefully inadequate. Somewhat reluctantly we accepted the notion that some sort of massive land connections had existed. Even as recently as 1971, however, in a lengthy book I had written over the preceding five years, I included only the slightest hints that continental mobility might hold the answers to the biogeographic dilemmas I addressed. The answers were, however, really there to be grasped at the time. At length, even the most adamant diehards gave in and accepted the newer theories. Older concepts based on a rigid uniformitarianism were gone!

From application of the new ways of looking at the rocks of the continents and oceans there emerged evidence that, by the beginning of the Permian some 280 million years ago, the great land masses of the earth were coming together to form what came to be a single continent. The continent, called Pangea, was completed during the Triassic. The land masses that are now North America and Europe met earlier to form Laurasia, one of the cores of Pangea (Figure 3). The single continent of Pangea lay astride the equator, which then passed through what is now central Texas and to the south of the latitude where Moscow now stands. It was around this equatorial zone of Laurasia that terrestrial life, plants, insects, arachnids, fresh water fishes, amphibians, and reptiles were concentrated. Beyond the continental limits a great ocean stretched from the west shore to the eastern margins of Laurasia. Much of the recorded marine life of the time was concentrated in shallow seas that overlay the extensive continental shelves and low areas of the continents.

Much later, during the middle and upper Mesozoic, Pangea began to fragment and the formation of the continents as we know them today began to take place.

Collisions of the once separated land masses that went to form Pangea resulted in the formation of persistent mountain ranges, among them the Hercynian-Appalachian chain between two ancient continents, Laurasia to the north and Gondwana to the south. The Ural Mountains developed along the eastern border of Laurasia as the continent moved, and later these mountains came to form much of the border between present day Europe and Asia as the Siberian continent merged with Laurasia (Figure 3).

During the Carboniferous period, continental ice sheets had formed in the Antarctic regions of Gondwana, leaving unmistakable traces in the rocks they overrode and the sediments they deposited. Around the equator the climates were little affected, but the rise and fall of the seas, likely due to melting and freezing of the ice masses, produced cycles of sedimentation during which formed the great deposits of the Carboniferous coal swamps. By early in the Permian these cycles of rising and falling sea levels had diminished and gradually disappeared.

Marine animals, among them trilobites, corals, clams, snails, nautilus-like cephalopods and one-celled foraminiferans, had flourished in shallow continental and shelf seas from near the beginning of the Paleozoic era. As Pangea developed, these seas were expelled from the continents and the living space of their inhabitants was drastically reduced. Coincidentally, and probably partially as the result of the expulsion of the seas, the

Figure 3. Pangea, the late Paleozoic and early Mesozoic continent, showing the approximate boundaries of present day continents. The southern block (Gondwana) included South America, Africa, Antarctica and Australia. The northern block (Laurasia) included North America, Europe and Asia.

marine animals underwent the greatest mass extinction recorded in geological history. Mass extinctions had happened before and were to happen again, as when the great dinosaurs died out, but never to the extent that occurred in the Permian. Animals and plants on land were affected as well, but less severely. Changes were, however, great enough to set the stage for a new expansion marked by the rise of the Mesozoic reptiles, including the dinosaurs, and the first birds, mammals and flowering plants.

Climates over the whole earth were severely modified by the formation of Pangea, increasingly so as the Permian progressed through its 50 million years. Early in the period the warm, lush equatorial climates of the Carboniferous coal measures began to yield to somewhat drier and more seasonal conditions. Desert climates gradually prevailed around the equator as the growing mountains blocked the easterly equatorial winds, and all traces of Permian terrestrial life in the North American equatorial region appear to have vanished. As the equatorial climates deteriorated, however, areas both to the north and south of the paleoequator appear to have become more suitable for life on land. It is from deposits formed in these latitudes, best known from the Soviet Union, South Africa and East Africa, that Upper Permian vertebrates are found.

Interpretations of shifting continents, Pangea, paleoclimates and distributions of plants and animals based on the "new geography" are very recent. They may again undergo radical changes in the future, but whatever happens they have radically modified the concepts of earth history in vogue when our work in the Permian began.

Our field studies at first followed directly in the footsteps of the earlier collectors in Texas, unaware for the most part, as they were, of the increasingly important finds being made in the Soviet Union.

The first collectors in Texas, beginning in the 1870s, were hardy men, no Jeeps or even Model A Fords, no bulldozers and not much store-bought food, but rather a team and buckboard, a rifle, some hardtack and a bag of salt. The problems of getting onto land where fossils occur, often difficult later, were more those of Indians than of ranchers and the Bureau of Land Management. Some 65 years of tramping the Texas red beds looking for fossils had convinced the various collectors by 1935 that those red beds which formed during increasingly hard times,

Figure 4. Texas, showing the location of the Lower and basal Upper Permian beds discussed in text. (Modified from Figure 1 in E. C. Olson, 1989, The Arroyo Formation (Leonardian: Lower Permian) and its Vertebrate Fossils. *Texas Memorial Museum Bulletin* 35.)

as desert conditions approached, would produce no fossils. The picture emerged of a great desert devoid of all vertebrate life after the time of deposition of the early Clear Fork, the Arroyo formation (Figure 1).

After having had some fair success in collecting in the well known Texas localities, I began to wonder, as others had before, if there might not be fossils in the higher beds (Figures 4, 5). The rocks didn't look much different; they just had not yielded any fossils. Young and improperly indoctrinated, I did not take the "great desert" idea too seriously. So we tried going higher, which meant going west from the standard sites, for the beds dipped gently to the west—about 50 feet to the mile—so that successively younger beds were exposed westward across the section. This dip, along with the simple structure of the beds and only a very modest increase in altitude, assured that higher beds were exposed in that direction. This venture paid off surprisingly well, taking us well into the base of the Upper Permian. But we, too, ran into a great desert with beds of gypsum, sands and mud, a miserable mix for producing fossils. Land animals almost surely were living somewhere in North America, but we have not been able to find their remains even through weeks spent searching in all of the areas where we felt they might be preserved. Perhaps someone, sometime, somewhere will find them and our "great American desert," too, will become a myth.

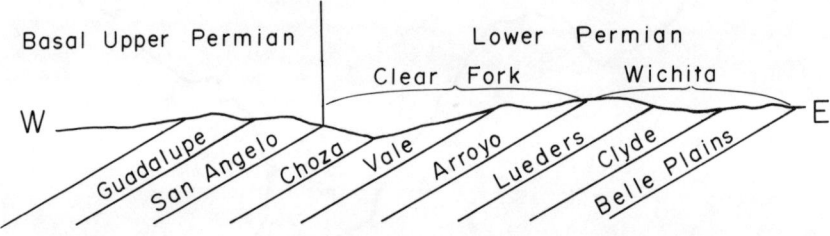

Figure 5. An east to west section across part of the Permian of Texas showing the formations treated in the text. The westerly dip of the beds is exaggerated, being only about 50 feet to the mile. A plan view of the formations is shown in Figure 6.

2

The Arroyo Formation, Coffee Creek and Ernest

When I first began to work in the Permian of Texas in 1936, I went to the red beds of the Arroyo formation north of Lake Kemp in north central Texas (Figure 6)—right where Paul Miller (Figure 7) suggested I should start. My ventures started humbly and naively, sort of learning on the job, on the Waggoner Ranch. Some miles in from the highway, I was down on my hands and knees scraping at a poorly preserved skull of the extinct reptile *Dimetrodon*. The skull was buried in some hard, dry red clay, coming near the surface some 275 million years after the demise and burial of the animal it represented.

I was intent on my digging, head down, when I looked up to see Ernest Cruthirds sitting high on a big gray horse which somehow or other had been maneuvered close by without my companion Father Rigney or I hearing it. From my vantage point Ernest looked big, tough and forboding.

"What the goddamned fuckin' hell you think you doin' here?" didn't help much, as I stood up, reaching hastily to pull out my permit from Robert L. Moore, ranch executive. Father Rigney, unaccustomed to the language, which we later found was just Ernest's way, sort of cringed in the red dust, hoping, I expect, that it all would go away.

"Oh," said Ernest, after reading the memo from Moore, "You'll be one of them goddamned bone diggers?" This with a big, snaggled toothed grin, which made me feel a bit better.

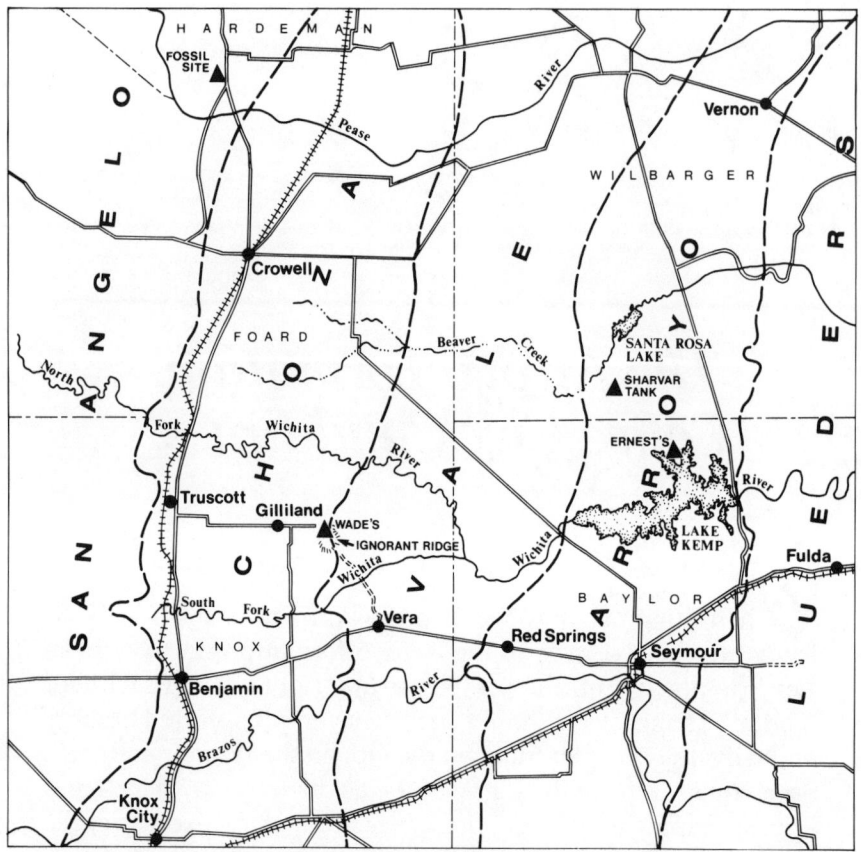

Figure 6. Detail of that part of Texas treated in the text, showing the locations of towns, rivers, camps and sites as discussed in the text. Note in particular the geological formations: (from east to west) Lueders, Arroyo, Vale, Choza and San Angelo. The boundaries of geological formations are approximate.

"Well," he went on, "you'll better get your water from my cistern. This stuff'll give you the shits." We already knew that.

Ernest (Figure 7) wasn't gentle in his talk and, it turned out, he was nearly tongue-tied if he couldn't intersperse his nouns and verbs with strong, expletive adjectives. Along with his talk, his deep red, sunburned face, capped by the cowboy's sallow-faced forehead, his dirty blue shirt and heavy leather chaps and spurs were enough to frighten any city bred Texas neophyte. Ernest was a real, genuine cowboy, a dying breed, and like his kind, having made his entrance and scared us enough, he turned his horse and rode away.

Figure 7. Left: Ernest Cruthirds on his porch, 1950. **Right:** Paul Miller, about 1940, in the Sand Hills of Nebraska.

We—Father Rigney and I—were pretty miserable. I hadn't quite learned to cope with the Texas breaks and weather, and Father Rigney really never did. It was mid-summer, crackling dry and everything in the country seemed to be trying to bite, sting or stick us. The deep red, warm water of West Coffee Creek wasn't much good for washing or bathing and least of all for drinking. So Ernest's cistern water—even with its frogs, live and dead, and mosquito larvae—was a godsend.

We stayed a few more days in our miserable camp. About the third day after Ernest's visit, the late "Hod" Sawin, in later years to become a dean at the University of Houston, showed up with his gang from Harvard. They arrived at our camp about noon, after being lost during the morning, and never really got over the shock of finding my assistant lying in the meager shade of a mesquite reading his Bible, snapping off an occasional insect pest. Bone digging and religion just didn't seem to mix to us young, budding scientists who more often than not proclaimed our adherence to an atheistic "faith" which I understood better then than I do now. An even worse hazard arose later when, in Seymour, Texas, a filling station attendant with our best interests at heart tried to "fix" Father Rigney and me up with dates with a couple of the town girls. I first here began to appreciate the real and practical problems of mixing science and religion. Father Rigney, I must say, in proper "philosopher's" fashion bore this mixing well and steadfastly followed his course to the doctor's degree in science while retaining all the grace of his position in the church.

Ernest was usually away when we went for water, so I didn't see much of him until later. After a few more weeks the summer was gone and we headed out.

I didn't see Ernest again until 1938, when I was back once more on the Waggoner Ranch. He was still living in his two room wooden house a little east of the old Seymour-Vernon road. When Lake Kemp (Figure 6), constructed for a reservoir, was filled about two decades before this, the old road had been abandoned and a new one put in over the dam about five miles east. Some eight miles of gradually decaying red clay road led from the Flippen Creek gate of the ranch to Ernest's place. As the old bridges went out and crossings washed away, new ways around were found, so that by the mid-1940s all but the first part of the original road was gone.

There was another way in, but even with this not many people came to see Ernest. Occasionally his mother stayed with him, as did one or another of his seven sisters, but mostly he was alone, the proper stereotype of the lonely but proud and self-sufficient cowboy. He "rode line" for Waggoner, getting his keep and a small wage for tending a pasture of well over 50,000 acres. The ranch as a whole covered 550,000 acres; it was split up into pastures ranging up to 100,000 acres in area. Ernest checked and repaired fences, tended cattle tanks (earthfill-dammed water holes), saw to the health of the cattle, checked on strangers and poachers on the land and, when the "wagon" and its roundup crew arrived, aided in directing their activities in his bailiwick.

This pattern is pretty much gone now. The roundup hands live in town and come out to the ranch by pickup. Helicopters flush the cattle and the cowboys guide the animals into pens and corrals. Then the cattle go to feedlots and off to market. It's more efficient, but it is nice to have lived a few days with the older way in its last days. Even then, the wagon was a specially equipped truck, although a team of six mules still was in service on the Four 6s Ranch in King County to the west.

When I returned to Texas in 1938, I began what was to become a long association with the ranch and came to know Ernest much better. Contrary to my first impression, he was not a big man, standing about 5 feet 6 or 7 inches. This hard, cussing cowboy turned out to be rather shy and somewhat childlike in his curiosity about what was going on in his world. His profanity and vulgarity were a ruse, a coverup, used when he first met someone. What I came to know of him came from several seasons spent with him in his Coffee Creek rider's camp and from yarns, with various versions, from the people of Seymour, Texas. The two pictures were very different.

Ernest grew up locally and had a less than admirable reputation in Seymour and the surrounding area. He was with cattle most of his life and seemed to delight in being the hell raising, shoot-em-up cowboy when he got to town on a spree with beer and spirits to cloak the shyness. A night of cooling in a local jail usually did the trick, after which Ernest was again free to go back to his cattle.

One time, so one version of the story goes, Ernest and a young friend of similar persuasions managed to rope a bull by the horns and drag it through the railroad station of Fulda, terrifying the people on benches waiting for the train. In one door they went and out the other, the bull bellowing and Ernest and his friend hollering and cussing and away they careened down the railroad tracks as the trained puffed in. Now, there are other versions of this story, some with two stations and some denying that Fulda was anything but a cattle loading station. True or false, the story was already legend in 1938. It was one side of Ernest. The last time I heard it, in 1979, it was much improved. That's a side of Texas.

Part of it was that Ernest liked his beer and liquor well, but couldn't handle them. In the later 1930s Seymour was dry, but to the northeast, towards Wichita Falls, beer was still to be had. One time we drove Ernest to Wichita Falls in our Model A. We needed gypsum plaster to block out fossils, but every time we passed a beer joint, and there were many, Ernest would plead, "Doc, I'm getting might dry." So we pulled over. He got very incensed that I would not drink one-for-one with him, but I had to drive. Finally we were getting lunch in Wichita Falls, and

Ernest, now in great shape, was roundly proclaiming about his exploits in the whorehouses of the town. In spite of his colorful protests, we were unceremoniously tossed out of the restaurant while Ernest and all of his 5 feet 6 inches wanted to beat up everyone inside and on the street. But this wasn't Ernest, only the man who couldn't handle liquor well.

On the ranch it was different. Ernest, like the other line riders, knew his land perfectly and had an uncanny knowledge of who was on it, where they were and what they were doing. We never needed to fear getting lost, for if we didn't know where we were, he did and, if necessary, would ride out and bring us in. During the springs or summers of 1938, 1939, 1940 and—after WW II—until about 1950, I spent several weeks each year on Waggoner lands working out of Ernest's Coffee Creek house. Ernest's house lay in the middle of some of the most fossiliferous areas of the Arroyo formation (see Figures 1 and 6). Ernie DuBois, Bill Read, Roy Reinhart—some of the graduate students who assisted me with this field research—shared his hospitality. I came to learn something about ranching, ranches and ranchers, and developed an immense respect for Ernest, the old reprobate by town standards. Most of what I picked up from Ernest came during long evenings while sitting, waiting for the day's heat to drop off, and swapping yarns.

To get along anywhere comfortably, of course, it is necessary to understand the local mores, at least well enough to avoid violating tacit taboos. In north central Texas it is important to understand the local variety of Texan speech, not so much the pronunciation as the pacing, the lack of emphasis on anything important and avoidance of anything that smacks of being an outlander.

Ernest, Ernie DuBois and I were sitting at the evening table having supper one evening early in the summer of 1938. Ernie and I were talking about the day's work in rapid-fire, Yankee style. Ernest, after a long silence, let us know in no uncertain terms, by Texas standards, that he didn't think we were behaving in a kindly fashion.

"Doc," he said softly, "I know the words you're saying, but I don't understand none of it." Back to his plate of beans and hide.

Gradually the slow, unhurried pace, the art of indirect palaver, and local colloquialisms began to sink in. At the table, "I wouldn't care for any," was understood to mean, "I don't want any more, and don't urge me!" When trying to get on a man's land to collect, "I don't guess you better," turned out to be a flat "No!" and no sense arguing any more. Hanging over the "bob-wire" fence for half an hour or so and talking was friendly. "Had much rain?" long pause; "Looks like you might make twenty bushels" (of wheat per acre); "Cattle looks good, but how'd them white-faced Angus get in there?" and finally, "Mind if I walk over them breaks back there?" all went a long ways further than an abrupt, "Can I look on your land for fossils?" Slowly we came to understand this genuine way of good neighborliness.

Here we were, of course, wanting to go on land, at least to walk and look it over, to go through gates, under and over fences and finally, perhaps, to take our vehicle in and dig out fossils. We were of no earthly good to the owners and especially suspect on big spreads which were subject to all sorts of hunting, poaching and "pickup" cattle rustling. Land had an almost sacred meaning.

Gradually all of this became a part of us. I think that the attempt to conform to the local mores helped develop friendship and understanding. One side of living with Ernest that I, or my field partners, never really mastered, however, was a taste for his culinary efforts. We all ate together, of course. Cooking was no problem, for Ernest had a butane stove and even a gas-operated refrigerator big enough for a hotel. Except when the wind blew out the flame, the refrigerator was great, although it was used mostly to store beer when someone brought some in. Ernest's favorite dish was not good beef, as cowboy lore would have it, but a somewhat uncertain stew. This was usually put together on a Sunday, simmered all day and night, and eaten during much of the week. The base of the stew was a five or six pound slab of fat side-bacon. Some lard was added. A couple of cut up cabbages, several large onions, carrots, and a half peck of potatoes for a thickener went in. Add water until the five gallon container was nearly full, put in some salt and pepper, garlic if handy, and simmer to serve. Cook up a batch of unleavened "hoe-cake" and spoon the stew over it, garnishing with small pieces of cut green onions. Salt heavily and eat. Hot pepper sauce helps.

A near crisis in the art of stew making hit in 1939. This year there was a plague of the giant, nearly wingless grasshoppers, locally called Mormon crickets. By the time we arrived, the grasshoppers had eaten the tops of all the onions and Ernest could not seem to find the onion bulbs in the hard red clay where they had grown. He asked us to dig for them with our small hand-picks, so-called Marsh picks after their inventor, the famous fossil hunter. So we did, we found them, and we hung them to dry in his shed, called the garage—but car-less. The Mormon crickets were not slowed by the garage, however, and they ate all the onions there—and if one cricket got hurt, they ate him too. Thereafter, that year, the stew was made with store-bought onions, and Ernest felt it was never really the same.

Some days, when near the end of the week and the stew was running low, the diet was hoe-cake and hide and beans. This mess was kept in a glass jar in the refrigerator, which jelled the copious amounts of lard used in cooking. Whether stew or hide and beans, lard was an important ingredient. We were working enough and sweating enough that we could take care of the fat; the problem was that Ernest had a half full, five gallon tin of rancid lard that was his pride and joy. Everything had to be cooked in it. It seemed to agree with him for all the awful things that even good lard is supposed to do.

Each evening there was a polite battle between Ernest and me to get to the kitchen first. I had a five pound tin of good lard. If I won, Ernest let me cook Yankee food—a steak, pork chops or chicken, potatoes and vegetables. If we had beef, Ernest would eat it, although reluctantly. He would insist on making hoe-cake. Chickens were only for special chicken fries, which required lots of beer to ward off the bad effects. But pork, sausage or anything but beef was not on his menu. He would turn to the always present hide and beans or stew. We even tried tricks to get away from the rancid lard. We drained the coffee can on the stove and put in good lard and capped it with a bit from the main supply. Ernest never said a word, but the next day ours would be gone and his would be back.

Ernest seemed to thrive on his lard and lived to a reasonable age. I was on an ulcer diet for nine months in 1939–1940, but probably just had a ruined set of cells lining the stomach. Tobacco was another thing. Ernest, like many cowboys, smoked Bull Durham, rolled carefully with two hands, rather than the one-handed technique of the movie hero. I tried Bull for a summer and decided that it must be a matter of deprivation that prompted the Bull mystique. So, one summer when we left, I brought Ernest a carton of Camels, for which he thanked me warmly, in lieu of saying goodbye, which he never did. His eyes were always a little sad as we left him to his life alone, but he never said come again, or when we arrived, asked us to stay. This was all done by indirection over the course of a half day or so. We all knew how it would come out from the start, but some of my assistants got a bit restive as a seemingly homeless evening approached.

Next spring when we came back, I saw the carton of Camels on the shelf, unopened. Ernest really liked Bull Durham! So I "ran out" of cigarettes and smoked up the Camels. They were a true smoker's cigarette. I think it was partly the Bull that began to get to his lungs. In the mid- and late 1940s he would get up in the dark, at 4:00 A. M., like he did with the "wagon" in the early days. He'd heat the coffee and sit and roll cigarettes, while racked with a hard, dry cough until dawn.

The cowboys I know all had a weakness for patent medicines. Ernest continually sent out for asthma cures.

"Hey, Doc," he'd say, "that goddamned stuff done me some good." But it never did.

Medicine wasn't all. One spring, about 1946 if I recall the year correctly, an Avon lady, of unusual enterprise, "made" the ranch, getting even to quite inaccessible places. According to Ernest, most of the cowboys stocked up. On what I am not quite sure. I know that Ernest got some deodorant with a powerful odor. He always smelled of good sweat, leather and horses, plus a touch of onion or garlic. I assumed he bathed, although I never caught him at it. There was a washtub out back for this purpose. We swam in the muddy waters of Lake Kemp, but no self-respecting cowboy would do this. One day I came in and found Ernest rubbing something under his arms.

"Ernest," I asked, "you got some sort of problem there, under your arm?" Not good Texan, but not bad if said slowly.

"Oh," he said a bit shyly, "some goddamned lady was selling this goddamned stuff to us. I didn't get none of her fancy perfumes or soap, but this here cream makes you stink good."

3

Eyes West—to Younger Beds

By 1946, our work around Ernest was beginning to play out, and I had begun to toy with the idea that we should move our fossil hunting to areas that were too far to walk to and too rough to drive back and forth every day. This would be new territory to us but one that Paul Miller and George Sternberg had worked over along the Wichita River before Lake Kemp had been put in in 1917. As things turned out, 1946 was a good time to think of a new base camp, for this was the year that Ernest's house burned down. We did spend part of that season in his pasture, but we also explored the area to the west with his help.

The house burned while he was riding fence and, of course, no one was around to see it go or try to stop it. His mother had been living with him for a time and, now and then, one or another of his sisters would help out. All had left, but not before tidying up and putting up curtains. Ernest, used to some help, apparently was careless and while he was away the wind likely blew the curtains into the flame of his gas refrigerator and started the fire.

I didn't know about the house until I came into his pasture about 11:30 A.M., just down from Chicago. I found Ernest and another cowboy sleeping in the garage on makeshift beds. It was 100 degrees plus and they were sleeping it off. Ernest rolled over when I punched him and exposed a completely, momentarily horrifying, red side, undershirt, pants, skin and all. The dye on the coverlet was not sweat proof!

He looked downright awful and his half-opened eyes pretty well matched his red half. As usual, no greeting, just as if I had been there all along.

"Where's your house?" I asked.

"Burned down," he muttered. "Took my goddamned new suit. Doc," he continued, "you goin' to town?"

I had just come from town, but said I could.

"Go to the drug store," he managed. "See the owner, tell him Ernest's sick, needs some medicine." He sat up, scrawled out a "check" on a piece of brown paper, putting on the name of the bank and town, and scribbled the amount. Signing it, he gave it to me to pay for the medicine. I drove back to town fully expecting to pay for the medicine out of my own pocket. The county was dry by then, so I was curious about what I might get. It turned out to be a pint of under-the-counter gin and the druggist happily took the check, which I gave him hesitantly.

About three o'clock I got back to Ernest's place. His friend had somehow gotten away. Ernest uncapped the gin and took about half of it in one swig. He downed the pint in a few gulps, while I cringed. A little water and he was up and about, feeling better. All I could think was that these cowboys are rough and tough.

Living alone and riding his pasture, Ernest, of course, knew his range well. He was curious about everything on the range and had developed a crude sense of biology that stood him in good stead. Cattle, coyotes, rabbits and all such mammals he understood.

"Like people," he told me once, "when they fuck, they get young'uns."

Snakes, lizards or scorpions didn't intrigue him much, except that some stung or bit. The striped grass lizard, *Cnemidophorous*, was a scorpion, and the scorpion a stingin' lizard. Rattlesnakes were just for killing. Birds came closer, especially chickens. He kept a few hens for eggs. Their sex habits were worthy of interest.

I got this story from Professor Romer, who dropped in on Ernest many times and knew him well.

"Professor," Ernest asked, "do you know about animals and birds?"

Romer, a professor of biology and anatomy at Harvard, allowed as how he did.

"Well," Ernest went on, "I've been watching my goddamned chickens for a long time now. Did you know a hen don't have to be fucked to lay an egg?"

A sound observation, Romer granted.

Another time, I was at the cabin when Ernest came in with a squawking, jumping bag, containing two roosters.

"Doc," he called over, "you see them hens there? I got two het up roosters in this tote-sack. They ain't never seed a hen and them hens ain't never seed a rooster. Let's see what happens."

So he dumped out the roosters, whose feet began going before they hit the ground. No niceties, no courtship gestures, just plain rape. Before they knew what was going on the unsuspecting hens were tagged and lit out, with their own indignant squawks, for the mesquite with the roosters in hot pursuit. By dusk three hens and two roosters were in the chicken roost, seemingly happy with affairs, with the roosters on the top rung and the hens on the bottom. Two hens never came back; probably the coyotes got them.

Like most persons who live on the land, Ernest was a conservationist, taking what he needed, but not destroying the land's capacity to provide for all. But when rules seemed unreasonable, they were to be circumvented. The Waggoner Ranch is a game preserve. Rabbits were fair game, but the big "blue" quail and wild turkeys were rigorously protected. We once even received a caution for shooting bullfrogs for food. The ranch had its own game warden. Occasionally Ernest wanted quail and it seemed right to him that he should trap them for his own use. Rabbits were beneath his dignity. Turkeys, too, were a delicacy, especially if forbidden.

One morning, while we were still there, Ernest left on his dappled mare with his shotgun, rather than his old rifle which he usually carried strapped onto his saddle. He told me that he had three shells and that he was going to get a turkey from the wild flock to the north of his cabin. I had seen him hit a rabbit from a moving car with his rifle, but also seen him miss a knothole target in the yard from 50 feet while standing still. After that he adjusted his sight with a hammer! According to his philosophy there was nothing wrong with us having a turkey or two, but I was a bit more dubious, or apprehensive, after the bullfrog incident.

About four o'clock, when the hot, drying day had brought us in early, Ernest came back with one shotgun shell left and three turkeys, ranging from about six to ten pounds. In a few minutes they were cleaned and in the monstrous refrigerator. Feathers, of course, were all over the yard and would fly up in our faces when the sandy, gusty wind blew.

In about an hour, sure enough, the game warden rode up. He patrolled the whole ranch, the whole half million acres, and my education in Texas ways was never up to knowing just how he spotted illegal activities. Helicopters were still a few years away. Whenever we would be out of our usual haunts he would seem to show up, just checking on who we were. It was more than coincidence, I am sure, that he rode up just when he did, neither too soon nor too late.

"Ernest," he started out, "I got some reports of shooting up north of here. You know anything?"

"Can't say I do," from Ernest.

The game warden, "You know that flock of turkeys up the way?"

"Up thar som'eres."

"Well, someone's been shooting at them. That's a jail offense."

"Should be," said Ernest.

All during this conversation, of course, gusts of wind were blowing turkey feathers all about, right in our faces.

"What's that, over there?" asked the warden pointing to a wooden apple box tilted up on a stick with a string attached to it.

"Rabbit trap," replied Ernest.

"Quail trap."

"Come on over," asked Ernest, moving toward the trap.

"See that fur?"

He pulled out some gray rabbit fur from among grains of corn. "See, the poor rabbit caught hisself on that nail in the box before I got him out. Can't get no quail near a box like that."

"Well, you know trapping quail means a fine and jail too," the warden told him.

"Yeah, I know, and a good thing too. Don't want no poachers on this ranch," came back Ernest, closing off the conversation.

Finally—it seemed like an hour to me—the warden got on

Figure 8. Camp at Sharvar Tank on the Waggoner Ranch, 1951. **Left to right:** Neil Tappen, a student of anthropology. Neil continued on to become a distinguished Professor of Anthropology at the University of Wisconsin at Milwaukee. Robert Bader, who after a career in zoology at the University of Florida, Gainesville, became Dean of the College of Liberal Arts in the University of Missouri, St. Louis, Missouri. Robert Sloan, who deserted his early love of invertebrates for a career as a vertebrate paleontologist and became a professor in the Department of Geology and Geophysics in the University of Minnesota at Minneapolis.

his horse and rode to the gate. While Ernest was opening it, the warden said,

"Ernest, if you hear any shooting, let me know."

"Shore will, got to keep them turkeys safe from them hunters."

The warden rode off.

Gradually, during 1946 and 1947, our work took us farther and farther west. We were trying to find fossils in beds younger than any that had yielded them earlier and this meant gradually moving away from Ernest's cabin. First we went to Sharvar Tank (Figure 8), some 18 miles west from the main ranch headquarters, called Sachuista, which lies on the Vernon-Seymour road.

We camped there for parts of two or three seasons using the water from the tank for everything. Bryan Patterson, then of Chicago, later Harvard, visited us and brought along the owner of the property he was working on farther east.

This old gentleman, wise in the ways of the area, refused to drink from the tank. He noted an over-ripe steer carcass a couple of hundred yards away and the buzzards that ate on it sometimes flew over the tank losing their cargo. The tank was big, and cattle drank from it and took care of their other needs in it, but none of this bothered us outlanders. We drank the stuff and felt fine. Not the old gentleman. He drank wine which he had brought. Wise as he was, and I am sure he was really right, Bryan and his group had to leave after two days because the old man had a serious case of the "runners."

The last time we camped at Sharvar Tank, Ralph Johnson and I set up camp and then noted how low the water was. This was during one of the periodic droughts in the region. The drinking did not look too good so we made coffee with the water. Coffee never did much to change the color of the red water, and we would use only red water because clear water meant a high content of gypsum, second cousin to epsom salts. This time, as usual, the color did not change, but the taste was terrible. We found out how bad boiled green algae could be. Not only was that batch bad, but the coffee pot was ruined, for no end of boiling and scouring could get rid of that foul taste. We changed plans and began to go six miles out to a line camp and haul our water. We could still squat down to bathe, but the value of the procedure was dubious.

Talking over "going west" on the ranch with Ernest turned us to what some people called Ignorant Ridge (Figure 9). I don't know why this name came to be, and did not find anyone who did. Some didn't even believe that the ridge was called Ignorant Ridge. It lay at the western edge of the Waggoner property in Knox County, some seven sand-and-red-mud miles north of Vera and about five miles of sticky black mud and sand from Gilliland. There were about 20 feet of Pleistocene gravel, sand and black soil on top of the Permian beds there and this made for good farming on the gumbo and an abundance of well water. Ernest had told me to go over there and, at the top of the hill, to look up Ab Covington, who was keeping the west pasture for Waggoner.

Figure 9. Top: Camp on Ignorant Ridge. Students slave and bosses supervise or sleep. Bob Miller was good at dishes but better at the mathematics of evolution. At this time a graduate student, he went through all the stages up the ladder at the University of Chicago, becoming a Professor in Geophysical Sciences. Having struggled to teach me math, Bob left fossils and evolution to become an expert in near shore wave processes and sediments. **Bottom:** Camp at Ignorant Ridge, our home for several years. **Left to right:** Russ Guthrie, outdoorsman, who with his Ph. D. from Chicago went from the Permian to become an outstanding student of Pleistocene mammals, their taphonomy and biogeography from his base at the University of Alaska. The "boss" in the seat of honor and leisure. Ted Cavender, fossil and recent fish expert who, after his Ph.D., continued his research at the University of Michigan and Ohio State University. One of the best men with a shovel and pick on our crews.

"You go see old Ab and tell him Ernest sent you. You say to him, 'How are you, you old son of a bitch.' He'll know I sent you."

So I did, and Ab did. I was ready to duck, for Ernest had often told me how this gentle greeting to strangers had got him in trouble. But Ab understood and somewhat later we set up camp on Ignorant Ridge about 200 yards from the house and with a rich supply of water. The water was a problem for a little while until our systems adjusted to its moderate gypsum content, but thereafter nothing tasted quite so good. The site was beautiful, overlooking the broad valley of the South Fork of the Wichita River, a richly green valley where the luxurious grass and mesquite fought a never ending battle against each other. The mesquite always won unless controlled by the ranchers. Cedars—really junipers—topped the hills and on even the hottest nights a breeze welled up from the valley. With this move, I didn't see Ernest for long stays and sort of lost touch with him.

One year, soon after, I found that Ernest had been moved to the Flippen Creek gate of the ranch, where Byrdie Fergusen, who kept the gate, took care of his failing health. By summer he had moved to a house in Seymour. The Waggoner estate, like many large ranches, was a highly paternalistic organization, taking care of its people, so Ernest was in no want. Ab Covington also had been moved to Seymour, so he and Ernest had each other for company.

During that summer I went to Seymour to look up Ernest. He and Ab were at his place and we sat and swapped yarns for a time. Ab's old dog, Dan, one of several from the ranch, ambled up and flopped on the porch with a gasp of solid comfort. As usual in the late afternoon that time of year, clouds had been building up. They announced their presence with a ripping blast of thunder. Old Dan, contented no more, slunk into the house, moving faster than I thought he could.

"Old Dan sure git when it thundered," Ernest allowed.

"Didn't used to," from Ab, "but an old mare I had attracted lightnin'. Everytime I rode her and a cloud come up, sure enough bolts would start hitting around us. Old Dan finally got so he wouldn't go out with that mare and me. Finally it got so he just wouldn't go out at all."

"Had a mare like that," mused Ernest. "Just seemed to draw a cloud. Had to git rid of her. Wonder if she ever got hit?"

"That was a while ago," Ernest went on. "Seems a long time back."

"Yup," said Ab, trying to get back in the swing of the talk.

"Hmm," said Ernest, and went on.

"Brings to mind one evening with a storm comin' on. We was a bunch of young bucks settin' on Mrs. Bates' porch, where we was boarding." Ernest began, keeping Ab out of the yarning. "We was drinkin' beer and every so often one or another would get up and piss off the porch. Out come Mrs. Bates and says 'You fellers quit pissin' off the porch.' So," went on Ernest, "we went into the yard and pissed on the porch."

His big, snaggle-tooth guffaw . . . .

"Pissed on the porch." Another guffaw. "Pissed on the porch. Ha, ha, ha, ha . . ., ha . . . ."

Next spring, I heard from Byrdie that Ernest had been very sick. He was back in Seymour, but still wasn't doing too well. His whole system was just running down, complicated by his asthma or, more likely, emphysema. The story I got is as follows.

In the Vernon hospital the doctors had pretty well given up on keeping him going and since he had a short time asked what he would like. Nothing he could ask for could do him much harm. Ernest rather weakly suggested he would like a good drink of whiskey. Just where to get it was a problem, for the area was dry.

At that time Robert Anderson, later a prominent member of the Eisenhower Administration, was Executive Manager of the Waggoner Ranch, which had its headquarters in Vernon. He was greatly liked and respected by the people of the ranch and of Vernon as well, and had aided in reducing the conflicts of the ranch and town that had long existed. So the hospital asked him if he could procure a bottle of whiskey, which he did. Ernest began to feel a little better and with a second bottle was well enough to leave the hospital for Seymour once more.

I rather imagine, knowing Ernest well, that the feeling that everything was over for him was as much to blame for his critical condition as his obviously bad health. He no longer had his horses, his morning Bull Durham and his coffee. He couldn't cuss and blaspheme as liberally as he used to, and the old red hill north of his old cabin, where he wanted to be buried, was far away. There was just nothing to do and his way of life had not developed any resources away from the land.

As soon as I could I went to Seymour, along with Nick Hotton, one of my graduate students, to see Ernest. He was in bed when we got there, about 3:30 in the afternoon. I suspect that he spent most of his time lying down. He never read much and, of course, there was no television. He started to get up, brightening, when we came in, but we just propped him up on a pillow.

"How ya doin?" I asked.

"Not too good, Doc. Not too good."

"Ernest," I said, "Byrdie's told me you have given up swearing and cussing. Is that right?"

"Smoking too," he replied. "Seems to get my lungs." Then with the old twinkle in his eyes. "I never did go much for this God stuff, but I'm goin' pretty soon. Don't mind, grant you, but figured I'd take no chances with cussin' so I quit."

"Must kinda hamper you, doesn't it," I quipped.

"Ain't no one to talk to, nothing to say, no breath to say it anyhow, so don't make no nevamind."

We chatted together for about an hour about old friends, Byrdie, his mother and sisters and my old field partners, Ernie, Bill, Roy, Mike and the lot. Where were they now and how had they done? Never figured Mike would get out of the bush, he was so short and meek, should have had a flag attached, and so on . . . .

I never saw Ernest again, for soon he passed on. He had always wanted to be buried in the red hill near the house on Coffee Creek by the Old Seymour Road. But, of course, it was not to be. Yet, when I go past the hill, as I do once in a while now, I feel some part of him has rubbed off on it, for it seems to have some of the gentleness and hell raising cussedness of this man resting in its bare red sandstone and clay.

4

Ignorant Ridge—the Vale and Choza Formations

The Ridge, Ab and Fossils

The southeast point of Ignorant Ridge stands over the magnificent valley of the South Fork of the Wichita River. As it turned out later, this was an excellent center from which to gain access to beds of the Vale and Choza formations which overlie the Arroyo. Many years before, in the 1870s, Charles Sternberg, a professional fossil hunter, had started exploring for fossils downstream along the South Fork. With his team and buckboard, he made his way to the confluence of the North and South forks of the Wichita River, passing the ridge along the way. Not until he reached the confluence (Figure 6) did he find any vertebrate remains. There he came upon a jaw, set with rows of teeth and quite unlike anything anyone had ever seen before. Its identity remained a mystery until into the late 1940s when a find by J. Willis Stovall in Oklahoma gave a clue to what it was. It proved to be an ancient reptile, about three feet long, to which the imposing name of *Labidosaurikos* was given.

A legend grew over the years that the beds in the area of the ridge, or anywhere west of Sternberg's find, had no fossils. So my plan to search them met with skepticism, justified enough. The late Ted White, an experienced Texas bone hunter and for years director of Dinosaur National Monument, hearing of my plans, wrote me, saying,

"Sternberg, Case and Miller went crazy trying to find bones in the 'Double Mountain.' Give it up or it will get you too."

Professor Romer, in his definitive monograph with Llewelyn Price on pelycosaurian reptiles had reiterated the general opinion that, with the end of the "Clear Fork" times, recorded about at the head of Lake Kemp (Figure 6), the continent of North America had become barren of vertebrate life. The great areas now in Texas, Oklahoma, New Mexico and Arizona were vast, desert wastelands around the saline seas of the Permian Basin of Texas. So around the ridge, to the east and well to the west lay the sands of a barren desert which gave way at length to the waters of a highly saline sea.

Still, during those long years of World War II, which I had passed in the unglamorous job of teaching maps and aerial photographs to an unending stream of GI's—Civil Affairs Training School officers destined for post-war Japanese Administration—and civilians, at the University of Chicago, I dreamed in off moments of getting back into the field and having a try at those higher beds. I just could not believe the whole continent had become a lifeless desert. I reasoned that any fossils that might be there must be in the sort of beds that earlier hunters would have tended to ignore, based on their work in lower horizons. Perhaps coarse, gravelly conglomerates might offer a chance. So I dug through aerial photographs, which were becoming available, looking for likely places and hit upon some areas of high relief north of Vera, Texas. Ignorant Ridge was part of this area and my chance camping there, on Ernest's suggestion, opened the way to developing the plan.

After the war, we started looking at once in some coarse stream channel deposits that in part were holding up the terrain and producing some of the relief. Within an hour we found fossil vertebrates. With these initial finds and my "brilliant" deductions it became possible to open up this whole area and to fill out a nearly complete sequence of fossil vertebrates for the whole of the Lower Permian, adding the top layers to the extensive earlier work of many paleontologists in the lower beds.

A triumph of the scientific method—inferences from data, prediction, testing and sweet success? Not really. The high relief seen on the aerial photographs was due largely to a thick layer of coarse Pleistocene rock, about two million years old, not 260 million year old Permian deposits as I had thought from the photos. The coarse channel deposits of the Permian did, it is true, form some of the relief and hunting had to be done on side hills, where no casual hunter, passing through, would look. So much, at least, was OK. In this area, the fossils were in fact just in the channel deposits, but in other places they were not. The reasons others had not found them was just that they were rare and scattered, and that the myth—stemming from Sternberg's trip—dimmed enthusiasm for any intensive search. The urge to "always go over the next hill," even though it is 110 degrees in the shade at four in the afternoon, was missing. So, as usual in science, sound deduction or intuition, luck and coincidence were the important ingredients, coupled with faith, hard work and a salting of stubbornness.

The first summer we spent on the ridge was flushed with the sense of adventure. As usual, land problems soon turned up. After a couple of weeks of digging I found out that we were not on Waggoner's land, for which we had a permit, but on Nichols' Ranch. The ranch foreman in Red Springs said he could not give us permission to dig. We would have to get that from Mr. Nichols, in Dallas. I always disliked asking permission over the telephone, for it is too easy to say no. So, the foreman let us finish what we were doing and, that winter, I made contact with Mr. Nichols, who graciously gave us permission for the next year. So often, we would get onto the wrong ranch, not knowing it, and get "caught." Rarely did we get permanently run off, but sometimes it took a bit of diplomacy and a lot of time.

Ab, as Ernest had said, turned out to be a genial man, waiting his retirement, which was to come in a few months. He didn't work very hard and came down to the tent almost every evening with Dan, the lightning-shy dog, and sometimes Curley, who stayed when Ab left. Ab loved to talk. One evening, after about a week of small talk, he got around to asking,

"What are you fellers looking for anyhow?"

"Fossil reptiles," I replied. "You know, like lizards and snakes."

"In them red rocks?" he asked.

"Yes," I went on, "been there some 250 million years or so." This didn't impress him at all.

"Thought so," Ab mused. "About 25 years back, an old

feller come by while I was still at Hog Creek Camp, over east. Name of Miller. Distinguished sort of man with a pointed gray beard. Had a funny way of talking. Dr. Miller as I recall. He was huntin' for some fossil lizard, he said, dimeterdum seems to me he called it. Something like that anyhow. Yeah. Dimeterdum. Guess he found one, too, but I never did see him again."

Paul Miller had worked with me for a number of years and, before that, with other Texas hunters, Samuel Williston and Alfred Romer. About 1924 or thereabouts, he had taken Romer on a trip to the Permian to show him some of his fossil localities. He was, indeed, among other things, hunting for the fossil reptile *Dimetrodon*—a long spined animal, several feet in length (Figure 10). *Dimetrodon* was the old dependable reptile seen in early science fiction movies as the prototype of the horrible, spine-backed monster, usually portrayed by a tired varanid or iguanid lizard with a floppy web taped to its back.

That Ab would remember Paul Miller and "dimeterdum" was unusual, for most of the ranchers and town people who would visit our digs, interested while they were there, usually didn't grasp what we were doing or remember what we were finding. They even seemed a bit suspicious that there wasn't anything there and that we were really looking for "minerals"— gold or uranium, as the times dictated.

One day, however, two cowboys who had seen us around with the wagon rode up and handed me several shiny, sharp, conical teeth.

"These here teeth what you lookin' for?" one asked.

"Yes," I said eagerly, forgetting my Texas manners. "Where'd you get them?"

These were teeth of *Dimetrodon* and they had been broken out of a jaw or skull, which I would like to have had.

"Wall," drawled the first cowboy, "let me see. You go down this dim road there about a mile, maybe a mile and a quarter, you see a red hill off to the west. Half way up is some sand rock and sticking out is this thing with the teeth."

Down the road a mile or a mile and a quarter, of course, were some dozen red hills, each just about like the others. One stood out to the cowboys, but not to me. *Dimetrodon* stayed undisturbed in his final resting place, now gone with wind and weather. But the cowboys had found something and this was about the only time this had ever happened to me. Mostly it

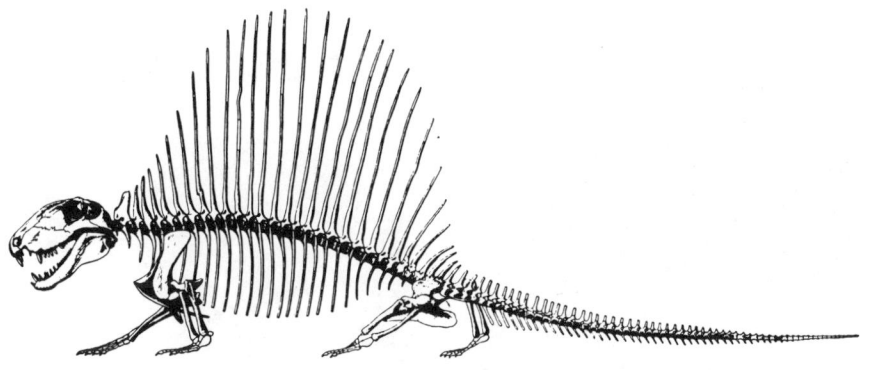

Figure 10. *Dimetrodon*, the "famous" long spined, primitive mammal-like reptile from the Lower Permian of Texas and adjacent states. Approximate length 5 feet. (After Romer and Price, *Review of the Pelycosauria*, 1940.)

was, "I been riding them breaks for 20 years and never seen nothing like that."

Ab remembered and had an idea of what we were after. This was before television had come along and done its immense, if less than well recognized, job of mass education. Now, with travel series, animal series and documentaries on fossils, coupled with readings in the Reader's Digest, everyone seems to know about fossils and what we are doing. The last 40 or so years have seen an immense change.

Next year Ab was gone, retired to Seymour, and Wade and his wife Maude had taken over the house on the ridge and the west pasture. Like Ab, they welcomed us and were of immense help to our studies and general well being.

The Ridge and Wade

Whatever a wise man may be, Wade was one of a handful that I have ever been sure of. I learned to trust his judgment as I came to know him better. One or two times he overestimated my abilities, but this was more my fault than his. Wade (Figure 11), in his days of riding line for Waggoner, was true to the image of the straight sitting, tall, taciturn cowboy. His bronzed face was wrinkled with the weather and, like Ernest, his forehead had the ivory cast of a confirmed hat wearer. He spoke

Figure 11. Left: Wade Barker, my favorite cowboy and guide to the lands and ways of Texas, in heavy fence rider's gear with his favorite horse and dog. Photograph taken on the mud road down from Ignorant Ridge in the late 1960s. **Right:** Wade and Maude Barker at the annual Santa Rosa Roundup staged by the Waggoner Estate of Vernon, Texas.

with a sharp, tangy drawl that overrode the range winds and the cacophony of bellowing cattle.

Wade, to me at first, was turned over, high heeled, tooled riding boots, leather chaps, rusty spurs, a blue jacket, hard hands and blue-gray eyes that squinted against the sun and dust, in Texas ranch style, and a slow smile. Cowboy life, with cattle and range critters, seems to merge a hard exterior roughness with a gentle nature, layered with an elemental, sensitive humor. Wade had it, so did Ernest, and never did I meet a cowboy of the legendary Black Bart stereotype.

Maude, Wade's wife, taught school, for a time in Gilliland and later in Crowell, some distance to the northwest. She and Wade lived in the small house at the top of a steep, mud-gravel road leading up from the valley. Going to our camp we passed through a gate by the house. We also got our water from a tap near the house. Usually, if someone was around, I would stop and chat, gradually coming to know Maude and Wade better as time passed. Sometimes, late in the evening, I, alone or with some of my party, would drop in and talk.

"Well," Wade would start, "Where'd ya go today?"

"Oh, down past Perch Tank," I might reply. "Got in the rough breaks near the east gate of the trap."

"Get through the gate all right?" Wade would ask with a bit of grin.

"You make those gates too high and too tight for me, Wade," I would joke, for I am but 5 feet 6 inches and Wade was better than 6 feet and remarkably strong. The retaining loops on his barbed wire fence gate were so high I had trouble with them.

"Takes a man to get through them," he would say.

"It's OK if you're sitting on a horse," I would come back.

And so on for a half an hour or so.

The ranch was fairly well crossed with mud roads and the roads all went somewhere—to a trap, a tank, a flat spot where branding was done—but most seemed to do so in a peculiarly round-about way. The reason was simple. The first wagon or car across, often following cow paths, made a track. Coming back, the tracks formed a guide through the grass and mesquite, so they made the start of the road. The next vehicle followed the trail, and so on until the tracks were worn. Horses and cattle sometimes kept the roads open. When a gully appeared across the road, the next passerby would detour around and start a new passage. At times the old road was lost and the new one consisted mostly of these "go-rounds."

Off-road driving, while necessary, was hazardous. Holes and bad crossings were one thing, but the worst problems were cactus and mesquite thorns which gave an average of three or four flats a day. This was before the days of steel belted radials, heavy mud grippers, and tubeless tires. Later, straight seismograph roads were cut, easy to drive, but rarely "getting anywhere."

When I first got to Ignorant Ridge I was driving a four door, 1929 Model A sedan (Figure 12). It was a fine field car and, like its ancestor, the "T," could be fixed easily. One time Nick Hotton and I were a long way out when the "A" quit. It turned out that the little buffer that broke the contact of the points when the cam rotated had broken off. We used Duco cement to put

Figure 12. Getting around over the years. **Upper left:** Our Model A, a fine field car, at a river ford in 1949. Bob Miller and Nick Hotton, now Curator in Vertebrate Paleontology at the US National Museum of Natural History, Washington, DC. After his years working toward the Ph.D., Nick spent seven years teaching gross anatomy at the University of Kansas at Lawrence and then got back to his first love, vertebrate fossils. Nick was a constant companion and coworker during our early exploratory forays into the higher Permian of Texas. **Upper right:** A 1951 Ford pickup, solid but skittery in mud. Ralph Johnson, another of my student teachers and one who made me conscious of field ecology, is seated on bank. From herps, especially snakes, to invertebrate ecology and geology, Ralph's ventures into shallow marine invertebrate assemblages set the stage for the studies of many to come. After becoming a professor at the University of Chicago, Ralph was

appointed chairman of the Department of Geophysical Sciences, an onerous but rewarding task. On the side, he was also a two term Mayor of Park Forest, a suburb on the south side of Chicago. Both Bob Miller and Ralph Johnson passed away, much too early, at the peaks of their careers. **Lower left:** Our chevy pickup, even more skittery than the Ford, following a muddy road in Texas. **Lower right:** A CJ 5 Jeep, which would go almost anywhere, aided by a winch on the front. Everyone worked hard—for subsistence, not salary. With Jeep and scoop to help, Robin Zawacki and Jim Edwards managed the drag while taking blocks from the Kahn Quarry in 1970. Robin strayed from geology and paleontology into computer science. Jim received his Ph.D.; working under David Wake of the University of California at Berkeley. After productive scientific work on amphibian locomotion, Jim went to the National Science Foundation, from where he aided his colleagues with their many financial problems.

the buffer back on the points. The car ran fine for two days until we got in to get some new points.

The car, parked on a high flat, did startle a cowboy named Tony. He found us later, for we were in his pasture.

"That old 'A' yours?" he asked.

"Yeah," I replied.

"Sure is strewed out," he continued. "Reckon you'll get it back again?"

"Hope so," I answered.

We were about 15 miles out from the nearest town and in trying to find out what the trouble was, after first looking too hastily at the points, we took off the carburetor, coil, and various other parts. They were lying about, with the points out to dry the Duco, when Tony found the car. Later we had a 1933 V8 Ford, then 1951 and 1966 pickups, and finally a CJ 5 Jeep. Except for the Jeep, the Model A was the best for off road—especially wet day—driving. Not much comfort, but it got there and back.

Wade proved to be an excellent guide in getting us about the 100,000 acre pasture he rode. He knew every blade of grass and every rattlesnake on the place, or so it seemed. Sometimes he suggested places we might go, having a sense of the kind of rocks we were looking for. One day he put us onto one of our best sites.

"Doc," he asked, "have you ever been over by those green sand rocks near the north-south fence? Looks like it might be a good place to hunt."

"No," I replied, having no idea what he was talking about but believing he did.

"Wall," he drawled, "you go in past Perch Tank, through the trap and the east gate (that high gate again). Keep on for about four miles, you'll come to an old well, dried up now. Kinda wind up the hill northeast from the well on a dim road to the top of the ridge. You'll drop down east to the fence, and drive north a half mile. Then you got to walk about half a mile, up, down, up and then down along the fence. That's the place."

Wade had consummate faith in our driving. We got to the well, but the "dim road" had gone so we just plugged along up hill and then east and, by good luck, did hit the fence at about the right spot. This was a rich site, but there was no way to get even the "A" into it. So we carted about several hundred pounds of rock and bones on our backs, on several trips—up, down, up, down to the "A."

Another time Wade had an idea of a good place, but couldn't figure how to tell us to get there, so he went with us. We turned and twisted up, down and around, and finally he said we would have to walk. I looked at Wade's rolled over, high heeled boots and thought to myself, "He'll never get very far." Cowboys don't like to walk, but you can't sell them short. This one, at least, blistered our feet and taxed our wind while I was praying we would never find anything wherever it was we were going, for we would never get it out. We did, but Wade came to the rescue and figured a way to get the car fairly close in.

One time later, Wade did overestimate our abilities. Ralph Johnson and I followed his directions, now with the 1951 Ford pickup, which was skitterish, and managed a flat tire half way up a loose, gravel hill, with a steep slope to the right of the car. So, flat tire and all, we had to reach the top, without flipping over. Ralph, 6 feet 3 inches tall and 200 pounds pushed mightily from the downside of the car and finally, in jumps and starts, we made it and changed the flat. This was more or less in a day's work, but when we got back there was Wade, waiting.

"Had a little trouble, did you?" he asked.

"Yeah," I said, still disgruntled by the whole affair.

Then Wade proceeded to outline to us in detail what had happened, what we had done, which tire was flat, who had done what and why we had missed the path he had outlined for us. He must have trailed our tracks by horse, knowing our tread. I always scoffed at the old stories of Indian guides and trailers who could practically know the life history of a party after following the horse's footprints for a ways. No more.

If a strange car or strange horse came on his land, Wade knew it. If a fence was down, or damaged, he knew how it happened and whether it was caused by a cow, horse or man. If some stranger had opened a gate, even though he had also closed it, Wade knew. If it was going to be a bad day for rattlesnakes, he told us so. His ability to recognize detail totally astonished me and how he did it escaped me. It was not only on his land. He confided to me one evening that he liked western movies but had to quit going because they were too inaccurate, not the plots but the props. The wagons might have the wrong wheels, the gun might be a few years too soon, or the particular hats were not worn then in the part of the country. I still don't know how he knew all of this, but I guess he did.

A good many years later, after visiting Central Asia and pondering on how the widespread empires of Ghengis Khan and Tamerlane could have been held together, I said to Professor Efremov, another of my wise men, that I couldn't see how the "war lords" could do it. How could they control such vast expanses without telegraph or telephones? His answer took me back to Wade.

"Only a big city man like you couldn't understand the 'long ear'," he said. "It's a wisdom of the land you will never know."

I never fully fathomed exactly what went into this "wisdom of the land" that I saw in Efremov, who had spent some years in Asia, or in Wade. It is passed down in legends from all continents. We city folks become adroit at dodging automobiles, we survive driving freeways, mostly, and our nervous systems stand up to the roar of passing trucks and trains and even the tensions of crowded airports. As a countrified urbanite, as I became during my many years in the field, I found that there are some senses so subtle that they never developed in me and that seem almost to smack of ESP. Perhaps they really do.

Unlike Ernest, Wade rarely swore or cursed around our group and never did at all when Maude was present. What he did elsewhere, I don't know, but I expect he could get off a fair selection of expletives under the right circumstances. He was always gently spoken with us, in sense although not necessarily in tone. One day we came back to camp to find Maude waiting for us. She wanted to tell us that Wade was in the hospital in Seymour and that she would be spending the night there with him.

"He came into the house about 2:30," she explained. "His face was all dirt and blood, his left arm was dripping blood and he was carrying his hat full of chicken eggs. 'Take 'em,' he said. His face was a blank stare. I asked what happened," she went on.

"'Goddamned if I know,' Wade said."

"I knew he was hurt," Maude said, "for he never did swear in front of me."

What had happened, it turned out, although he never remembered it, was that Wade, coming in early, decided to drive a loose heifer across the road into his horse pasture. Somehow, he didn't know just how, his horse shied or slipped and threw him so that his leg became entangled with his saddle rope. The horse spooked and dragged Wade some 50 yards. During this time he was slashing at the rope but mostly lacerating his left arm. Finally, freed, he walked back to the house, stopped to collect the eggs in his hat from several nests, and came in to see Maude. He was completely unaware of what was going on, having sustained a moderate concussion.

Getting thrown, breaking bones and having concussions are standard cowboy hazards. Getting dragged seems to carry a particular stigma. But if you cut yourself free, that's not so bad. At least this is how I interpreted it. Late that afternoon, Tony, another cowboy from up the way, came by and he and I retraced the steps of the accident in detail. Wade had been on his feet, behind the horse, for about 20 yards, his heels dug in and his spurs raking the ground. At that point he tripped and, with his free hand, he took his knife, a small one, and started to slash at the rope. With the jerking and rolling he hit his arm more than his rope.

Now, the critical point was, had he cut the rope or did it just break. We examined the part of the rope that Wade had left in the barn while he was collecting eggs. Tony asked me.

"Yes," I said, "I'm sure he did. There are good cut strands all the way through. I'll tell Maude."

"Oh," Maude said, "Wade will feel so much better when I tell him."

5

Land, Weather and Dogs

Ranchers and Land

One of the most time-consuming parts of fossil hunting, as any bone digger knows, is getting onto the lands for prospecting. In Texas we were not bothered by federal laws of the Bureau of Land Management or Indian Reservations, for we were digging on private lands. For a while, Texas did put in some collecting laws, but they were soon modified and, in effect, dropped. But wherever there are large spreads, as in ranch country, land tends to become sacred and going on it without permission is taking a chance—not nowadays of being shot (I only looked down the barrel of one shotgun) but of being "kicked off" and unable to continue the work already started.

The Waggoner managers, with their half million acre spread, were always gracious and, except for a few years when ownership problems cropped up, issued permits for legitimate studies on their lands. The permit, however, is a must. Most ranchers, if approached reasonably, are similarly gracious and helpful. The main frustration comes in trying to find the owners, for it sometimes entails a lot of running around over a period of several days. Then, after all of that, it may not work out.

Some land north of Benjamin along the Wichita River looked worth studying for fossils (Figure 13). The formation exposed in that area was clearly Choza and we had not yet

Figure 13. The many faces of Texas red beds. **Top:** The valley of the South Wichita River looking southeast from the camp at Ignorant Ridge. **Center:** Flood plain, over-bank and channel-fill deposits in Wilbarger County, north of camp on Ignorant Ridge. **Bottom:** Outcrops of lower San Angelo formation beds capping evaporite deposits of the Choza formation north of Crowell, Texas.

learned that the even, alternating red and green layers of that formation were formed in shallow, salty waters and lacked any fossils. The land was owned by an 83 year old doctor who lived in Benjamin. At Wade's suggestion, I went to talk with him. Wade knew him well.

"He's a funny old boy," Wade had said. "If you get him right, he'll say all right. If not, he'll likely say no."

So armed, I went to see the doctor and found him home. He was cordial enough, at least not hostile, as I gradually worked around to my wish to get on his land. He was way ahead of me, but he listened without saying much. When I finally told him I would like to get on his land, he didn't change his expression but just said, quietly, "I don't guess you better."

This sounded hopeful to me, so I explained a little more what I was doing, that it was scientific, not commercial, and that his land was in a specially important place.

"I know what you're doing," he replied, "but I don't guess you better."

"I know cattle, close gates, don't set fires and don't even smoke," I came back.

"I know," he replied again, "but I just don't guess you better." This time with a little more emphasis.

Soon I left, no further along then when I started. That evening I told Wade about the experience. He laughed and told me that when I heard, "I don't guess you better," I was through and should have quit right then. The doctor did not want anyone on his land. No reason. He just didn't want it. It was his land and that was that.

One time later, near San Angelo, Texas, came a reminder. I stopped one Saturday afternoon to ask about getting into some rugged hills where the San Angelo formation was exposed. First, I had gone to a house west of the road. It turned out to be the hands' house and they spoke only Spanish. They let me know that the owner was over east. That I went to "help's quarters" first may not have helped in what happened. Anyhow, the rancher came to the door, in his undershirt and drooping pants, beer can in hand. The television was blasting out a baseball game, so I went right to the point. So did he.

"That there land's mine, them gates is locked and going to stay locked. Ain't nobody gettin' on that land."

Quickly this time, that was that. Some years later Mr. Carey, in Taylor County, Texas, was different. When I asked him about my party looking over his breaks, he told me, "No." My wife and Rick Lassen were in the Jeep and could not see, they said later, why I didn't get in and leave. The straight, flat "no" somehow seemed to me not really so flat. It turned out, after more talk, that he had a "tank" full of fish and didn't want any fisherman. The main things that turned it were his old dog and a car that happened to go down the gravel road in front of his house too fast, kicking up a mess of dust.

"No good driving like that," I said.

"Killed my old dog the other day, one jest like'em," Mr. Carey said.

"Boy, that's no good," I sympathized.

"Old feller came dragging his rear end, all stove in, lay down and died."

"Worse than losing a friend. What was his name?"

"Jim. None like him, not now anyhow.

"If you want to go through that gate there, you can drive way back. Look out for the rattlesnakes, place is full of them."

Wade had some of this same sense of property and the inviolability of the land. Anything he had was yours, provided he gave it to you voluntarily or you asked. Down in the gully, falling apart was an old, discarded feed trough. One day while I was away one of my party had seen it and taken some broken planks to put around the bottom of our tent which was blowing and ripping. Wade was hurt by this, maybe a bit mad, but if so he didn't show it. How could someone take something without asking?

So I told the fellows to put it back, which they did, but without any understanding of why. After all, the wood was just rotting away. Wade, of course, forgave their ignorance about the proper ways of life.

"He's a good boy, but he just don't know about property, being raised in the city," Wade commented to me.

Wade had much the same attitude toward the people north of the river in Oklahoma. One day we had driven to near the Red River and were standing on the sands looking across to Oklahoma. The low, muddy central river channel passed into coarse grass, mesquite, "salt cedars" or tamarisk and then into a backing of oaks on both sides.

"Looks over there just the same as here," Wade mused, "but you know, the people sure are different."

Of course, this is true. The Oklahomans are very easy to know and not very suspicious of motives, except in the far western ranch country. The people of north central Texas, to me at least, were slower to accept strangers and waited to be shown what the motives back of their presence really were. It took about three seasons really to be accepted into one small town, but then acceptance was complete and impellingly gracious.

Wade, in keeping with general custom, accepted the graduate students who showed up with me, because they were with me. He sized them up quickly and usually correctly.

"Too much mother," he said of one. "Too small for this country, ought to put a flag on him," of another, sounding like Ernest. "He'll be fine when he gets over being scared of the outdoors," referring to one to whom field work was all brand new. "Never need to worry about him, he acts like a Texan," and so on.

Most of the young men enjoyed going up to the house and talking with Wade and Maude in the evening, and Wade never gave them any indication of how he felt about them one way or another. We talked mostly trivia, about coyotes, dogs, rattlesnakes and weather. Sometimes the fascinating history of the land came up and his comments were a lesson in lived history. Army life and Wade's assignment to take care of military dogs occupied many sessions. Wade said the army couldn't figure what to do with him, an old cowhand, so they gave him charge of the animals, which turned out to be dogs. The boys and I, of course, just ate all of this up. But we never got very far below the surface.

The deeper side came out gradually in continued contact over the years. There never was any oral self-examination. Wade knew who he was and what he was, what he could do, and how the world ought to be, and saw no sense in telling anyone about it. If he had no use for you, it came out, but only the more sensitive persons would ever recognize it. The core of likes and dislikes, of right and wrong, all centered about the sacredness of property and reflected the attitudes of this whole part of the giant state.

Wade, of course, was in strong favor with all of his ranching associates and was often away helping some neighbor work his cattle. Even after retirement from Waggoner's, after 65, he was in constant demand as one of the rare practical cattlemen still around.

Before coming to Waggoner's, Wade had worked for Halsell's in Foard County. One day he suggested that Halsell's might be a good place to look for fossils. The location of the ranch, north across the North Fork of the Wichita River and to the west, suggested to me that the beds exposed there might be high in the Lower Permian section. This was just what we were looking for, so I was eager to see what might be there. Wade drove up with us in the old Model A, now on its last legs, and we found Glenn Halsell in. Wade had told me he had better do the talking because Glenn was kind of touchy about people on his land. Glenn didn't seem too enthusiastic but after a time he agreed.

"I'll tell the hands not to bother you," he said.

"Don't I need a written note?" I put in.

"I said I'd tell them, didn't I?" closed the conversation.

We didn't get off to a very good start. When we got out to the Model A, we noticed a strong smell of burnt insulation. The ammeter had shorted and wires had fused. The battery was dead. At least the car had not burned up. After I had wired a bypass of the meter, we pushed the "A" down a hill, trying to start it, but we ended up in a blind fence corner, still dead. Glenn, red-faced from pushing, took a dim view of the whole affair, and I was afraid we had blown the whole thing. Then I took the crank, thumb over the handle in "broken arm" style for leverage, and spun the old, four cylinder motor as fast as I could. The rings were weak and the valves somewhat corroded, so the compression was low. I must have worked up a slight charge, for on about the twentieth turn the motor caught. Nick Hotton, at the steering wheel, slapped the spark down and we were ready to roll. After dropping Glenn at the house, we headed back for the ridge.

Halsell's proved a good place to work. It was easy to get around by car. Fossils were rare, and fragmentary, but anything in these high beds was important. Glenn Halsell apparently did tell his men to let us alone, with an almost eerie result. No one ever waved as we went by, no one looked at this old wreck of a Model A struggling along over the breaks. It seemed to us that, as we went by, they always turned and looked the other way. After a while this began to get to Nick and me. It was like being in a room, part of a trio, where two talked as if a third was not there. After a while the odd man begins to wonder if in fact he is there. It must hit a person in solitary confinement this way.

We were having this sense of isolation, when one morning, after a heavy rain, we could not drive in and started to hike the three miles to where we had been working. As we took off down the muddy road we noticed a man, seemingly with a gun over his shoulder, a couple of hundred yards back. He kept coming along, about the same rate that we were walking, but when we looked around, we would see him duck out of sight. Maybe we were in trouble. Maybe the word who we were had not gotten around to everyone.

After a couple of miles of this, we decided to stop and have it out. As we watched, the man approached, ducking out of sight once or twice. When he got to us, I hastily explained what we were doing and that we had permission from Glenn Halsell. He allowed as to how he guessed that was all right, but it didn't mean much to him. He was just walking the pipe line, checking for gas leaks! His "gun" was a probe. Our paranoia thinned a bit. Still, for the whole following month on the ranch, no one else ever spoke to us or acknowledged that we existed.

Weather, Bone Diggers and Cars

North central Texas spring storms are awesome. Cold northern air, usually from the northwest, tumbles headlong into the mass of warm, moist air up from the Gulf of Mexico. Of an evening, a first alert may be a few faint flashes of lightning far on the northwest horizon. The next morning a hot brilliant sun mocks the omens of the night before. Then a rustling of grass and brush and a refreshing push of the leading edge of the cool air mass under the hot, humid blanket. The cooling is welcome for a moment, but from then on all sorts of things may happen.

Wade and Maude's ridge was a marvellous vantage point from which to follow these nature-made spectaculars (Figure 14). Up the course of the South Fork a red sand cloud may rise into a thousand-foot wall, pressing east at express train speed. Temperatures drop from 90 degrees or more to as low as 45 in what seems like a few minutes. Lightning begins to flash and thunder makes even the wind's roar wane. Rain and hail start pelting and, somewhere in all that fury, there is a tornado sweeping across the land.

Figure 14. Wade and Maude Barker's sturdy stone house on Ignorant Ridge. Photograph taken from our camp east of the house while watching a storm riding in from the northwest. The tall water tower withstood this cloud and many others.

The house on the ridge, like most in the region, had a storm cellar. Sometimes these doubled as potato cellars with a proud king snake in attendance to keep out the rats. The one on the ridge had only one purpose, protection. Perched high above it, not far from the house as well, was that immense, iron water tank which a squeaking windmill kept filled with the delicious, somewhat saline water from the underlying Pleistocene gravels. It always seemed to me that a big storm might topple the tank onto the house or the storm cellar, but it never did.

During the time that Wade was in the hospital with his concussion, cut arm and broken ribs, we were in our camp down the ridge from the house when one of the granddaddies of Texas storms came through. First, in the late afternoon, there was the harbinger cloud of sand rushing down the river bottom and, not far north of it, a deep black, sometimes red, funnel cloud tearing up brush and gravel. With a final explosion of a barn about three miles away, it rose and passed some 500 feet overhead.

By night the storm was in full bloom, with lightning bursts that crackled out in all directions across the full vault of the sky. The wind became stronger and the tent shuddered. For a time our car radio worked, reporting such disturbing items as a train derailment in southern Oklahoma, caused by a tornado. Soon a lightning charge, dancing along the overhead wires to the ranch house, jammed our radio's capacitors. All in all, it was bad and we were a little scared.

Finally, about 9:00 o'clock, we lay down on our cots to wait it out. About then I had my one and only "convincing" encounter with telekinesis. When the rains were especially bad, I had often kidded my assistants that they were not to worry, because if it became necessary, I would exert my will and stop the rain. They decided now was the time.

"Okay," they said, "let's see you stop this rain."

"It'll take a couple of minutes," I shouted back over the roar on the tent canvas. "Be real still so I can concentrate."

So I concentrated on the task of stopping the rain, putting out real strong "brain waves." It stopped! The silence was almost overpowering—no rain, no wind, no night birds and no frogs—and no one said anything.

"Hey guys," I said after about five minutes, "I'm pooped, can I let it go now?"

"Okay," they laughed, and the rain came pelting down.

Once I shot a bullfrog at about 50 feet, hitting it right behind the skull, in effect "pithing" it, as in a biology laboratory. The sights on the gun were way off and everyone cheered. I never tried it again. I never tried the rain again either, so to date I am one hundred percent effective in applied telekinetic "science."

The storm continued violently for most of the night, and it was hard to get to sleep. Every so often I would peer out around the tent to look at the house. Maude had the lights on and so I knew she had not gone to the storm cellar. It must be all right, I assumed, knowing that she had a radio and telephone. The night did pass, the tent held up, and by morning there was only a misty drizzle. I went up to get some water and Maude came out.

"Bad storm," she said, "Hope Wade's all right, my radio and phone have quit."

"We looked up to the house last night, trying to see if you were all right," I continued the conversation. "Decided as long as you hadn't gone to the storm cellar things were okay."

"Well now," Maude replied in her quiet way. "That's odd. When the cloud was real bad I was looking at the tent and figured it was all right when you-all hadn't gone to the cellar. When you came I was going to go."

The weather is always a problem in collecting fossils in the spring in north central Texas. Rains and red mud make it nearly impossible to navigate the roads, much less to walk with 10 pounds of sticky clay on each foot. Spring's virtue, however, is that it is not as hot as it is in the summer. One time we managed to get our 1951 pickup mired in an arroyo, about 10 miles off the paved road. A forecast of general mist turned out five inches of rain that came down in an hour or two. This was one time I saw a tornado from the bottom of the funnel. The heart, or eye, appeared immensely clear blue.

Wade was not too worried when we did not show up that night. He knew we had to cross a river that would have flooded and would wait for it to go down. What he didn't know was how tightly we got stuck. The 1951 pickup, as pickups tend to do, dug in where the Model A might have "floated" across. There was a bad arroyo, one that always worried me even when it was dry. Anxious to get out before the river came into flood, I tried it when it was running water. We didn't come close to making it and so began digging in the muck, mesquite and cedar at about 10:30 in the morning. What with a 200 pound chunk of sandy gypsum lodged under the wooden tail board, we dug until dark. The night, at 50 degrees and with a howling wind that shook the pickup, passed slowly. There were three of us in the tiny cab, tilted at about 15 degrees to the side. Still covered with wet mud and getting hungry, we started to dig again about 4:30 A.M.

By noon we had the machine free. I must hand it to my two assistants, Swanny Swanson and Dick Seltin, for they took it all in stride. As we came to the river, I was delighted to see the old bridge was still there and now clear of water. It was an odd affair, planks fastened onto two wire cables and about 30 feet across. It went out in one of the next big storms and the ranchers made a ford, which was much safer.

We got back to Wade's in about two hours and found that he had just called the sheriff to organize a posse to hunt for us. He'd been down to the river and seen that it had dropped. Guessing that we were in trouble he started an abortive search.

Secretly, I felt it was a shame we got back so soon, because not too often nowadays do a college professor and a couple of his students get rescued by a genuine, western sheriff's posse.

The Model A didn't get stuck very often, but it did go through points, plugs, condensers and coils in fairly short order. I think the cam on the distributor shaft had a sandpaper surface. On the way back from a foolhardy venture, everything happened to it at once in a hard rain storm. In one of his rare moments of giving questionable advice, Wade had said we could probably drive from the Waggoner Ranch headquarters, Sachuista, across country to his place. We had done about three-fourths of this from one end or the other, but wanted to see the country in between. It was important to know the full section.

Nick Hotton and I had gotten about 20 miles past Sachuista when a rancher met us and said we couldn't get through, and anyway it was going to rain. We went a bit farther and then declared "to hell with it" and turned back. Night began to come down so we pulled off the trail a ways and bedded down. It was a miserable evening with black clouds scudding overhead, marked by eerie orange patches that shown through from the high clouds, which we could see through occasional holes in the lower ones. Finally, the sun went all the way down. Rain spittered but never fell in any amount. Mosquitoes were there in droves, so we had to put our heads down inside the bags, sweltering. About 4:30 A.M. I finally dropped off.

"Hey, ya'll lost?" some voice shouted at us.

Sticking my head out of the bag, I saw a truck up on the trail, going out to tend herd.

"Hell no," I said ungraciously. "Trying to get some sleep."

Fully frustrated now, we got out of the sleeping bags and made our way back to the ranch and the road. By the time we reached Seymour, it was delightfully cool. By the time we reached Vera it was pouring, so we did not stop for gas, which we really needed. We took the new, paved road across the valley towards Gilliland, but two thirds of the way across the rain drowned out the "A" completely. About then some "fool," I thought, came up and blew his horn right behind us.

"Good God, that's it," I vented. "Some bastard wants us to move over!"

It turned out to be a highway truck checking the culverts. "Need a tow?" he asked.

We did, of course, so we hooked on for a hair-raising ride to Gilliland, getting pulled much faster than the "A" would go on its own, and on the bias. There at the filling station, we pulled out most of the electrical system and put it on the wood stove to dry. While we were sitting in the car, Roger came by. Oddly, I don't believe I ever knew his last name, although we often crossed his land going north for fossil digging. He was driving his wife—Billy, of course—back from the hospital where she had been healing broken ribs from a bout with a horse. He stopped his yellow Jeepster.

"Give you a push?" he yelled at me.

"No!" I yelled back.

But it did no good, for we went a hoop and holler down the road to the end of the blacktop.

"Won't start?" Roger called over.

"Hell, no," I replied irritated. "The goddamned coil and distributor are sitting on the stove back in the filling station."

"Oh," said Roger without a blink, "push you back." Which he did—backwards—for a half mile in a short wheelbase Model A, that backed badly at best.

We got the thing back together again, filled with gas and took off. We made it to Wade's, but would have done better had we not got about 10 percent water from the heavy rain.

"Still moving," was our only comment through hundreds of yards of hub deep water, on a sand base, and a hundred yards of the fine, black gumbo, also hub deep.

As we drove in, Wade was tending his horses.

"You boy's have some trouble?" he asked.

We told him about our harrowing adventure, ending up with our watered-gas trip from Gilliland.

"Old 'A'll always get you there," he commented.

6

The San Angelo Formation— and a Look Backwards

Robert Roth, of the Humble Oil Company in Wichita Falls, earlier had given me invaluable aid in understanding the stratigraphy of the Vale and Choza. In 1949 I stopped in to see him at his office, and we went over some of the broad problems of Permian stratigraphy. During this talk, he noted some beds in which Humble Oil field men had found what looked like fossil bone. They were in exposures on the north side of the Pease River in Hardeman County (Figure 6). This was really exciting, for the formation in question was the San Angelo and it was considerably younger than any in which we had found fossils. The tip paid off and began to change the whole direction of our studies, eventually leading to my studies in the Soviet Union. While this new phase of our research was developing, we kept at the lower beds as well. I also took the extensive collections we had from the Arroyo, Vale and Choza and began to synthesize the information.

The Animals of the Arroyo, Vale and Choza

We had fully expected to find distinctly new kinds of animals in the higher beds of the Vale and Choza, but in this we were mainly disappointed. A few new species did come to light, and to one group of primitive animals with the impressive name of captorhinomorphs we added two new genera. A new species of lungfish from Wade's "green sand rock" added to the picture. The changes, however, were subtle as I suppose we might have expected. About that time, my close associate Robert L. Miller and I were delving into the application of statistics to the study of fossils. I knew fossils and he knew statistics, and we traded information. Out of this came a much better appreciation of the meaning of small changes. In this perspective, the information from the higher beds, when integrated with that from the older Arroyo sediments, led to the idea of studying the changes of a whole fauna moving through time, not just its parts alone. The concept was formalized under the model of a *chronofauna*, a name suggested by Bryan Patterson. The central idea of the chronofauna was that at a given time an array of animal species formed an integrated trophic system with each species occupying a particular position in the food chain. This is a standard ecological proposition. Over time, the species composition of the network might change while the constituent roles remained essentially intact. Within this context, the evolution of the system as a whole and the evolution of the individual lineages can be studied in a realistic and interpretable context.

Everything was not, of course, straightforward. A few rare and strange fragments and skeletons of entirely new animals did turn up in beds of the Vale and Choza formations. They did not fit the patterns at all. Some years before, Paul Miller had obtained skeletons of three otherwise unknown animals from a small pocket near the Arroyo-Vale boundary. Near the top of the Vale we found two intertwined skeletons of an early herbivorous pelycosaur, *Casea*, closely related to one of Miller's animals. One small part of a similar animal turned up in the Choza. Nothing like them was found among the hundreds of other Vale-Choza specimens. At one site a palate with eight rows of teeth was unearthed; at another, part of a large forelimb; and so on. For the time being they seemed mostly destined to pose unanswerable questions, unless our model was completely wrong. Later, as our work carried to still higher beds, close relatives were found, living under very different environmental conditions. During Vale and Choza times another evolutionary pathway was going along, somewhere in the uplands, and our odd fossils appeared to have been washed into the low-lying beds where the two systems came close together.

By the time these puzzles had come to light, but had not been solved, we had pushed our wandering to the uppermost beds of the Choza formation and had encountered sediments formed on the shores of the great Permian inland sea, where red clay-shales and gypsum portrayed an arid climate, born out by the adjacent land sediments. Fossils were absent and the "great desert" seemed to have caught up with our efforts.

The Upper Permian San Angelo Formation

Capping the barren upper Clear Fork sediments, mainly west of the Benjamin-Crowell Highway (Figure 6), were gray-green sandstones and gravels. These beds formed low escarpments and looked to be very unlikely sources of vertebrate fossils (Figure 13). Even after we had opened up some good sites with partial skeletons showing, Professor Romer, who had come over for a visit, assured me, tongue in cheek, of course, that "there warn't no fossils in them beds." He wasn't too far from wrong, for these beds, which turned out to be the deposits of the San Angelo formation, were for the most part barren of fossils and those that did occur were in sediments very different from the ones of the Lower Permian. This difference threw our ways of hunting off until we became attuned to new ways of operating.

Everything we did find, however, was new and exciting. The old water-based ecosystem of the Lower Permian, now gone, was replaced by a full-blown terrestrial system dominated by large herbivores and a smattering of smaller predatory carnivores. While fascinating, another startling development emerged as the study went on. We had begun to overlap in time deposits in the Soviet Union, from which had come their earliest Upper Permian vertebrates. Some of their animals and ours proved to be strikingly similar, so much so that in one or two cases they might be considered to belong to the same species. We had begun to forge a link between the basal Upper Permian of the two continents, North America and Europe. The high Russian deposits overlapped those of South and East Africa in time and animals. The establishment of a fully documented evolutionary sequence from the lower to the uppermost Permian loomed as a possibility. This initiated the chain of events that eventually led me to Moscow.

We started our hunt in the exposures of the San Angelo along the Pease River near the Benjamin-Crowell Highway (Figure 6), as suggested by Robert Roth. This was near the northern end of the San Angelo exposures, which pass underground shortly before reaching the Texas-Oklahoma border. The southern end of the formation was about 200 miles southwest in the vicinity of the city of San Angelo, the place after which the formation was named. The southernmost fifth of the exposures baffled all our efforts to find fossils, but the remainder, up to the Pease River, yielded its treasures, if reluctantly. The sites were separated by river cuts, but on the whole, we were able to get fossils from a more or less continuous line of exposures that formed a band about 160 miles long and five to ten miles wide.

Luck, as usual, played a part in the work. In our first venture into the Pease River exposures, driven there by rains that made the muddier Choza beds unworkable, we turned up fossils after about an hour of hunting. Nick Hotton and I were scanning the rocks not far from the highway when Nick called out, "Hey Doc, I got something here." I was about 20 feet upslope and hurriedly skidded down to him. He did have something—some teeth of a fresh water shark. While I rooted around, taking them out, he went up to the level I had been working, and 10 feet from where he had stopped me, located part of a pair of jaws! The game of one-upmanship in hunting fossils is always played by prospectors, and here my graduate student, Nick, had one-upped me twice. I had it coming. However that was, the jaws he found were of some animal like none we or anybody else had ever seen before in North America. In the smug vernacular of the pompous of our profession, it was "new to science."

The luck was in getting such quick results, for it was not until a week of fruitless search had passed that we found anything else! If these first bones had not turned up, we might well have quit the San Angelo, considering it barren of vertebrate remains. More did turn up and gradually we extended our searches southwestward, to Little Croton Creek on the MacFayden Ranch; to the southwest of Benjamin, Texas; to the Alexander Ranch near Truscott; and to the Swenson properties not far from Aspermont (Figure 15). Finally, we filled in the middle by working north of the Benjamin-Guthrie Highway. This was mostly done from the camp at Wade's, but gradually the distances got too great for the old Model A Ford. We were able to

Figure 15. Vertebrate remains in plaster jackets along Little Croton Creek on the MacFayden Ranch, Knox County, Texas. My crew (left to right): Dick Konizeski, who went from digging and shovelling to become Professor of Forestry at the University of Montana, Missoula. Ernest Lundelius, a true Texan's Texan who, returning "home" after a stint to his Ph.D. at the University of Chicago, made great inroads into our knowledge of Pleistocene vertebrates and has become a "wheel" in geology at the University of Texas at Austin. Dick Beerbower, who early prodded me into the consideration of faunas as units, put his agile wits to solutions of intricate paleoecological problems and is now a Professor of Geology at the State University of New York, Binghamton.

get hold of a used 1951 Ford pickup truck, more roadable but not as agile in getting over the bad country as the "A." At length, the time spent in running the 35 to 50 miles from Wade's to the collecting sites began to be too great. Reluctantly, I decided to leave Ignorant Ridge and establish new headquarters farther west. For one season we used Truscott, Texas, as a base. Later, Benjamin, Texas, became the center of operations, and with the discovery of a rich deposit some six miles west of Benjamin, along the Benjamin-Crowell Highway and six miles north on mud roads, we set up a field camp at the site which served us for several seasons.

It was tough going in the San Angelo wherever we went, for fossils were mostly scrappy and were found almost always one at a time, not in concentrations. One summer, for example, we spent six weeks and came up with eight specimens, mostly scrappy. So, discoveries north of the Benjamin-Guthrie Highway became milestones. One find of particular importance was made by Jack Kahn, a graduate student in sedimentology, who was out with me in 1955 when we were beginning to explore the beds on the Driver Ranch. He'd been walking the stream cuts south of the North Fork of the Wichita River, while the rest of us were out in the flatter country. Along one arroyo, Jack discovered a layer of sediment about three feet thick, with a concentration of bones that stretched out for some 40 feet! This was what we had been wanting.

From 1955 to 1960 (Figures 16–18), we excavated this area, which I dubbed the Kahn Quarry. About 200 specimens came out of it, some complete skeletons but mostly partial skulls, jaws, limb bones, vertebrae and so on. To hasten the work, we hired a bulldozer four times and got a hold of a gasoline-fueled jackhammer. We made a big hole, finally about 100 yards wide and cut some 50 yards back into the bank, with 25 feet of overburden. From this quarry, and of course other places as well, came a wealth of material that, in some ways, was very much like materials of the same age in the Russian Permian and, in others, very different. It was much scrappier than we would have liked but nevertheless of great value in the reconstruction of the life of these times.

Dying Towns and Houses

Truscott

People in the small towns of Texas are hard to get to know, but once they are your friends, there is the basis for as fine a relationship as can be. One spring Ralph Johnson and I, following up on work at the Pease River and Little Croton Creek, decided to stay in the little hotel in Truscott, Texas, just west of the Benjamin-Crowell Highway. Truscott had gone far downhill over the last several years as railroad service diminished and automobiles and good roads shifted the populations to only a few of the many towns in the area. Truscott, once with a bank, hotel, stores and a grain elevator, had seen good days. When I was there, however, the hotel was just holding on and the bank had long since gone. The elevator stayed and later was enlarged.

Figure 16. Camp at Kahn Quarry. The quarry was named after its discoverer, Jack Kahn, sedimentologist. **Left to right:** Herb Barghusen, who retired in 1988 from the University of Illinois Medical School to follow his photographic bent. Bob DeMar, another Chicago Ph.D. and student of Permian amphibians and reptiles, is now professor at the University of Illinois, Circle Campus, Chicago. Matt Nitecki, my long term field partner and "camp boss," who explained me to the graduate student gangs and them to me. Moving from the Walker Museum collections to the Field Museum of Natural History after receiving his Ph.D. at the University of Chicago, Matt became Curator of Lower Invertebrates. Beginning in 1956, and for several years thereafter, the camp was spring or summer "home" for many other graduate students in paleozoology including Jim Hopson, (University of Chicago), Pierre Joelicour (University of Montreal), Dick Seltin (Michigan State University), Vernon Swanson (M.D.), Bob Hessler (University of California at La Jolla), Dave Simmons (bone histologist, Houston), and John Donahue (who left vertebrate paleontology for greener pastures).

The church stayed and was strong and the school was maintained—held onto like unyielding defiance of grim death itself.

Such towns found a parallel microcosm in deserted farm houses (Figure 19). An air of sad nostalgia and dreams realized, then lost, emanates from a stately crumbling house standing bleakly alone in a field with corn or cotton crowding its thresholds. The sagging shutters once kept out the hot sun and the cold winds of the northers. An added dormer tells of the growing up of the children and a good year with the crops. The

Figure 17. Excavating at Kahn Quarry. **Top:** Orville Gilpin of the Field Museum of Natural History excavating with a gasoline powered jackhammer. **Bottom:** Block containing a nearly complete skeleton of the "giant" Permian herbivorous reptile, *Cotylorhynchus hancocki*. The specimen removed from this block is shown in Figure 18.

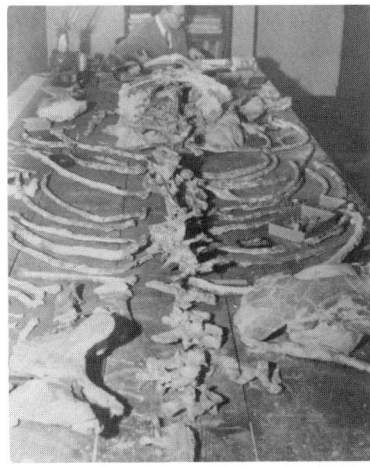 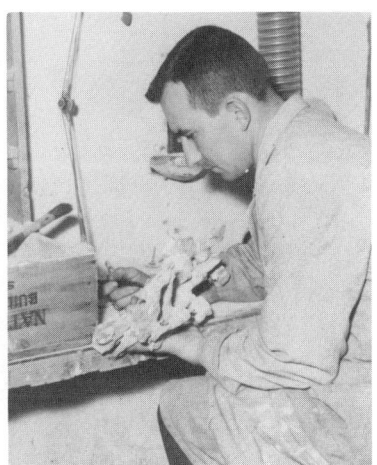

Figure 18. Left: A nearly complete specimen of *Cotylorhynchus hancocki* laid out on a table in my laboratory at the Field Museum of Natural History. **Right:** Matt Nitecki in the "sound proof room" at the Walker Museum, University of Chicago, working on vertebrae of the specimen.

tilting, wrap-around verandas once were set with chairs and swings where the cool of the evening was welcomed. Kids trudged down the faint, weedy driveways to hike the miles to school, kicking stones along the mud roads. They grew up, helping with the farm work, using new equipment and techniques, and putting new paint on the old house. Drought, autos, roads and wars, drawing off the young, left the old people on the land. At length, many who were left behind moved to town, close to friends and help. The fields of Texas and Oklahoma are dotted with these memories and, like the towns, they always start me dreaming of what must have been.

Once the hotel in Truscott had been the center for gatherings from the country side at the big family table—eat all you could for a dollar. It was still functioning in 1946. When Ralph and I moved in a few years later, the restaurant was still open, but mostly for short order service. We were then the only paying guests. The town was in the center of an area we wanted to work and we had the idea of putting together a book. The book never came off, but we both developed ideas that appeared separately in publications later on.

Figure 19. An old, abandoned farm house near Gilliland, Texas, and its weathered and degraded landscaping, telling a nostalgic story of what used to be.

Mrs. Jones ran the Truscott Hotel, keeping it open with the help of a few occasional customers such as us and as a meeting place offering coffee, coke and a juke box. Below our room was the restaurant and juke box, and we were regularly regaled by the semi-country western rock tune *I can handle that job all by myself*. At top volume!

Mrs. Jones knew everything about everyone around. Helping her was a delicate, small and very old black man called Goldie. He took our bags, bigger than he, and carried them upstairs, in spite of our protests. The relationship between Mrs. Jones and Goldie was fascinating. I have no idea how he felt about her for he never talked much except to comment on the weather. Mrs. Jones was ultra-talkative. One evening sitting in the cool in front of the hotel on an old wood bench, after Goldie had left to go to his room, she began to ramble.

"He's gone to his room to read his Bible," she explained. "He is such a good little man, so clean and so religious. He knows his scriptures by heart. Never knew a better man. Isn't it a shame he doesn't have a soul!"

There was no movie in Truscott, but Gary Cooper had left his mark. The kids unconsciously acted out the roles of their heroes, not only Cooper, but John Wayne, Alan Ladd, Jimmy Stewart and Henry Fonda. Late in the day, Ralph and I were having supper in the cafe when we heard the door open. But no one came in. We looked around and standing there was a tall, splinter thin, young man in his late teens, looking the place over. His flat, black hat shaded his slitted eyes as, head still, he shifted his glance back and forth over the room. Satisfied, he walked to the counter with slow, measured steps, eased himself down, pulled out his Bull Durham bag, and went to work on a one-hand roll of a cigarette. With a slight turn of his head, he looked at us with narrow eyes akimbo.

"What'll you have, Billy?" asked Mrs. Jones in her strident voice, cutting the image down to size.

Billy, pulling his tobacco pouch shut with his teeth, shifting his look quickly right and left, replied with a flat, drawled monosyllable, "Beans." High Noon at the lunch counter was never better done.

A few days later the big news broke. A wildcat well had come in near town. A wild flurry passed through Truscott, good times were coming back and the town would rise once more. Knox City, some 25 miles south, already was reviving with some gas and oil. Geologists from several oil companies came on the scene. Most of them, with big cars, stayed in Knox City, connected by a good road. One group of three men, however, stayed in the hotel. We got together for some "geologising" that evening, both figuring on picking the others' brains. We lost. You can't get much from nothing.

"What are we in here?" one asked. He meant, of course, what was the geology all about and, I thought, what rock formation was exposed.

"Well," I said, with a bit of caution, "those lower red beds with the gyp are Choza. You get some sands, kind of gray-green, and then some San Angelo. We are working higher, in the red shales of the Flower Pot (now grouped with the San Angelo)."

"No," came the dazed reply, "you know, what period?"

"Good God," I thought, but didn't say aloud. "You mean the Permian?"

"Yeah. This is Permian?"

These were oil company geologists? All I could figure is that the company must have wanted to get them out of the office. One was a graduate student, on summer pay, but the others, the boss at least, might have looked at a map. Anyhow, everyone in Texas, geologist or not, knows it is Permian in this area. A considerable part of the state forms the famous Permian Basin.

A day later the party boss asked me if there was any unit he could map on. The geology is excessively simple, beds flat and little faulted.

"Sure," I answered. "The top of the basal sandstones of the San Angelo is best." I explained in some detail where to find it.

The next day they were back. Couldn't find it. They found some orange peels where we had eaten lunch, about 100 feet up the hill from the mappable bed. Next day they found us in the field and were happy that they had found a good bed. I never did know if they realized it was the one I had told them about, maybe because I had another shock. Ralph and I had found a fair part of a skeleton, had dug around it, put a plaster and burlap jacket on it, and were hammering long cold chisels under it preliminary to turning it over. The geologists drove up.

"What you got?"

"Part of a big reptile. No head."

"Find it here?" from the youngest, a graduate student.

This stopped me cold. The block weighed around 600 pounds and was still attached to the ground by a three foot pedestal, giving it the look of a giant mushroom. The instrument man, who was pretty sharp, was embarrassed, and carried it off so I didn't have to answer.

All seemed to end well. We didn't see the party for a few days, but in the evening at the hotel, on the weekend, they were packed and leaving. The boss came over and thanked us for our help.

"I shouldn't tell you this," he said, looking around the deserted lunch room. "It's a company secret, but we got a structure!"

"Where," I asked, and after he had explained, "Great." I lied.

This country has a lot of gypsum under ground. It is soluble and as it leaches out the ground tends to heave or drop forming little basins and domes. Not much for oil prospects. I suppose it was a structure of this sort, but I doubt that the home office was overly excited. As to the wildcat, as so many do, after putting out a few hundred barrels of oil, it gracefully died and Truscott went back to its unrippled calm and slow decline of the weeks and years before.

The hotel didn't last the year, but the domino parlor, where a garage once thrived, stayed, and down the street the church stood firm. A new grain elevator went in near the tracks but the train stopped only to drop a car or two when there was a harvest transport to be done. It's all kind of sad.

Benjamin

I really came to like Benjamin, and in a way to love the little town. It lies on the north-south highway south of Truscott and centers where the east-west highway makes a stop-light cross roads. Ten miles south is Knox City. Benjamin still holds on, although its population is getting top heavy in age. It even picked up a little in the late 1960s and early 1970s. It had a school, several filling stations, a few stores and business offices, the court house and Hazel's Cafe. The last time I went through town, Hazel's was closed, with a "For Sale" sign on the door. Its passing was not matched by an opening of a new place.

It was in 1957 that we set up camp on Driver's lease northwest of town, where Jack Kahn had found the bone layer. For some five years that was our spring or summer home as we dug in our quarry. It was a marvelous site on which a whole generation of my graduate students cut their paleontological teeth. We opened it again in 1971, but the charm had gone and the old suspicions of ranchers, once dispelled, had all come back. It was a dreary anticlimax.

The people of Benjamin, numbering about 350, tolerated those odd strangers who holed up in that godforsaken back country. Mickey Driver, of the ranch, had a good idea of what we were doing. For all of his tough ranch manners, he was actually well educated. He tried to hide it. His aging mother and father came out to see the quarry once in a while. Mr. Driver was much impressed by the big hole, but allowed as to how his land was mostly holes anyhow. Spritely Mrs. Driver, all of 5 feet tall and 90 pounds, just looked on knowingly. While ranching went on, she took off, flying to Paris to see the world.

When we hired a bulldozer in our second season word got around that something really was going on out there. One day at the Texaco station a big, handsome man of 25 or so approached me. Right off he said he was a coach and teacher, in that order, at the school and could he bring his class out to see our dig. I said that would be fine but he would have to clear it with the Drivers. They didn't like anyone running back and forth over the five miles to our camp unless they knew who it was. The road went across the land of three different ranchers and there were five gates to open and close.

About 25 youngsters came out. They had fun looking at the bones, chasing lizards, looking for snakes and teasing each other. The teacher told me it was the first time he had seen a glimmer of interest in their eyes! He had come to coach, which he loved. He ended up with a physics class, which he hated. His young, pretty wife was pregnant and the small town closed in pretty tight. He had been a top basketball player at Texas A and M and, as he put it, a basketball bum. He still played the circuit in summer. Next year he was gone.

Later a church group came out. Evolution was a very touchy subject and pretty much taboo. I tried not to offend, for beliefs are delicate. But this group, on a Sunday, nearly stumped me. The reverend asked me to tell them what we were doing, which I did. They seemed mildly interested although a bit upset when I said the materials were about 250 million years old.

"Was that before the flood?" one elderly gentleman asked.

"Well . . .," I hesitated, while my graduate students looked to see how their prof would get out of that one.

"Well," again, "it was during a flood that these bones got put down, but not Noah's flood, a long time before it."

"Before!" exhaled the man in disbelief, "What flood before?"

"The San Angelo Flood," interjected the reverend with the grace of his office.

"Oh," said the man, relieved. But then, "Was it before or after the light?"

"Had to be after. These animals are green plants and they needed light," I said, getting back on the track.

I suppose I might have launched into an explanation of years, how to date rocks, how animals and plants developed (evolved) and how one might reconcile the Bible with geological events. But beliefs are important, and the myths that support them are the materials of human rationalization of the disturbing and unfathomable. I don't feel they should be tramped on too hard.

Once they had found the way to camp, some of the youngsters from school began to come out on their own and Mr. Driver didn't like it. Also they really cut into our working time. So I had the principal announce in school that they were not to come. I hated to do this, but otherwise we would have been thrown off the Driver property. But as the years went by I saw many of the youngsters in town, where we had become fixtures, even storing our gear in the loft of the service station. Late one afternoon, a delegation of three of the boys arrived at our camp, saying they had asked Mr. Driver and it was all right.

We talked and they hinted around that they had something on their minds. Finally one asked me if I would give the graduation talk at their eighth grade ceremony in a couple of weeks. I was mightily flattered and said yes, if it was all right with the principal. One of the boys was his son, trying hard to find a way into nuclear physics. I hope he made it. They carried my consent back over the 11 miles to town on their bicycles. I really couldn't refuse to talk—I was too moved by the invitation.

It was a fascinating ceremony. There were 13 boys and girls graduating, decked in their finest clothes and with caps. The gym was packed with 250 or more people, most of Benjamin. Prizes for the bests of the year were awarded. I had borrowed a coat from the principal, three sizes too large. First, to my amazement, I was given an award, for what I am not quite sure. A very pretty, blond teacher received her award as teacher of the year.

Then my "address." It lasted about 15 minutes and was more or less midwestern "corn." I ended up saying I might not be back too many times, but to remember that "big daddy from Chicago will be watching you and wishing you well." It seemed to go off well and when I dropped in to Hazel's a few years later, a gentleman come over to me.

"Ain't you the one who gave the talk at the school?" he asked. "That was a good one."

Ending the graduation was a talkfest outside of the school. The pretty teacher of the year came hurrying over to me. "You-all better change boxes with me before you go back out there," she said. "Take this one, it's got your shirt in it. The other's a frilly blouse for me. I don't think the boys would admire you in it."

We closed the quarry soon and opened it up just once more, as I said, after I had moved from the University of Chicago to UCLA. It wasn't the same. Benjamin was still there, but the young people and teachers were gone. The older youngsters were bussed to consolidated schools. For some years, I saw old friends as I dropped by the Texaco station in Benjamin and stopped to see Wade and Maude in Seymour. It was good to keep it alive, for there is so much to learn from these towns and their people that they must not just fade away. Who will teach humility to our young city people?

7

Eyes to the East

I began sending some of my publications to Russian paleontologists during and after World War II. About all I really knew about the paleontology of the Soviet Union was that there were extensive deposits of the Permian in the country and that fossils had come from them. There was a bit of relevant information in our literature, but not much. The language barrier was high and was compounded by the barriers of world turmoil and postwar misunderstandings. The last paleontologists from the United States to visit the Soviet Union had attended a geological congress in 1937. We had managed to isolate ourselves most effectively, so I was gratified when I began to receive publications from Moscow. Gradually I accumulated a fair number.

Most of the papers came from Professor Ivan A. Efremov. His work in the Russian Permian had paralleled mine in the Texas Permian during the 1940s in a rather remarkable way, but he was studying the Upper Permian and I the Lower. In several large volumes he combined his own work with that of his predecessors, bringing together for the first time comprehensive studies of the vertebrates and sediments of the Permian that stretched the length of the western flanks of the Ural Mountains.

As I began to cross from the Lower to the Upper Permian, with my early forays into the San Angelo, the Russian books and papers took on new interest. These publications had good pictures of animals, with their names in Latin, but the texts were incomprehensible to me.

Nothing can be more frustrating to a scholar than having a wealth of uninterpretable data piling up on his desk. I can imagine the tension that might accompany any message we might receive from other worlds, perhaps other systems, as it sits defying interpretation for years, driving the exobiologists and exolinguists crazy. In my case, however, there was a way around the problem—learn Russian. This task seems less formidable for German, French or Spanish, because at least the letters are similar to ours and don't look like caricatures of the backwards, upside down writing of juvenile graffiti. As usual, I took the direct route—I got an elementary text, studied it and then, with dictionary at hand, tried to translate the papers. Formal classes are, in the long run, better and quicker. My German, learned in class, has stuck; French, learned the other way, easy as it is to read, never really jelled. The same for Spanish, and so on. The problem is that, to study fossils, I really needed to be able to use these languages, and it seemed a chore at any given time to enroll in a class and take the time for disciplined study of the language I needed.

Never one to seek help, probably unwisely, I began my pursuit of Russian with *Russian Primer* by Agnes Jacques of Roosevelt College, and learned all about going to the country, having colds and finding the bathroom. Then I plunged laboriously into translating the scientific papers, word by word, phrase by phrase, and grammar be hanged. It nearly hanged me. I went at the job one hour a day, at noon, stubbornly and with no lunch companions. My verbalizations of the Cyrillic characters must have been something to curdle the ear. I did not know *how* bad, for I had no idea of how Russian might sound until, later, I ran into a cab driver in Moscow who had learned English this way. I had settled down for the trip when he began a torrent of completely incomprehensible articulations. I didn't know what language he was speaking.

"Po Ruski, pazhaluiste," I implored ("In Russian, please").

"No Ya gavaru po Anglisky," he replied, "ponymaytchi?" ("But I speak English, do you understand?").

"Ya ni ponymayu," I went on ("I don't understand").

"Vui Amerikanski?"

"Da," I allowed.

"Ya xachu gavarit po Angliski," he pleaded ("I want to speak English").

"Ne vozmozhno" ("Impossible"), I replied, and sat back and closed my ears mentally as I did often when the cacophony of incomprehensible conversation became phonetically unbearable. He rambled on in his "English."

A couple of years after I had been trying to master the Russian language, Professor McQuon, a linguist then at the University of Chicago, set up an oral-aural course in Russian complete with a Russian-speaking concert pianist to keep his students on the straight and narrow in pronunciation. This was a great help and we learned all sorts of phrases. Some such as "she's not a lady but a teacher" and "the outhouse is to the right" never did come in handy. Later, before I went to Russia, I had a tutor for a time and finally, several years later, I taught elementary Russian in an adult course in high school. I hesitated to tell my Russian friends about this, but it did help me immensely and during a subsequent visit to the Soviet Union, my friends there wanted to know who the teacher had been who had improved my Russian so much.

All of these studies were undertaken with a mind toward going to the Soviet Union. I approached Professor Romer, then at Harvard, about going in 1958, and we more or less decided on a trip together. I finally did go the next year, but his schedule spoiled his chances.

Preparations

During the time I was trying to find out something about the Russian Permian and trying to gain some proficiency in the Russian language, my correspondence with Professor Efremov flowered and we had become scientific friends. At the same time I was also in touch with Professor Orlov, then Director of the Paleontological Institute. Orlov was the one in the Soviet Union who had to make arrangements for my work in the Museum of the Paleontological Institute, where the collections I needed to study were located.

Two somewhat disparate sequences of events finally got me to Moscow, the cordial and helpful letters and aid from Professor Orlov and the odd, if perhaps typical, contacts with the Soviet Embassy in Washington, DC. I might have gone through the US-USSR exchange program of the National Academy of Sciences. Just why I didn't I am not sure, but I suppose it was just my old tendency, like the way I learned Russian, of doing things myself.

Following this bent toward personal independence, I made an appointment with Mr. Krylov, the Cultural Secretary of the USSR, to discuss my plans. My uneasiness about the Soviets was strong enough that I found passing through the doors of the embassy to keep my appointments a big adventure. The Secretary saw me at once and was most cordial. After a few minutes of conversation about my plans, he asked me abruptly, "What has your program to do with Morgan?"

"Really nothing," I replied truthfully enough. Professor Thomas Hunt Morgan was a Columbia University geneticist and one of the principal actors in the development of some basic concepts of genetics. He was the figurehead of "Morganism" when the ubiquitous "-ism" was needed for focus.

Probably I was overly suspicious, but I sensed a trap. More likely the Secretary was just trying to find something we could talk about. I did know a fair amount about Morgan and the monumental struggle over genetics that had raged in the Soviet Union, the so-called Lysenko affair. Politics and scientific incompetency had won out over science and knowledge, with the near demise of the vigorous program in genetics that had developed in the Soviet Union. Apparently, I passed the Secretary's test and we parted on cordial terms, with Mr. Krylov's request that I send him a detailed letter describing my plans. I did, but what ensued was protracted and only concluded just before I took off for Moscow. It is best portrayed in our correspondence.

Correspondence with Secretary Krylov and the Embassy

The agonies of getting visas are old hat, but the oddities of dealing with Soviet officialdom in 1958–1959—and now, too, I suspect—need a bit of elaboration. As noted, I saw Krylov and had a long and pleasant talk with him. As he requested, I sent a very complete curriculum vitae and plans for the trip. Mostly this was a one-way exercise. A few short letters give the flavor.

December 16th, 1958
Embassy of the Union of
Soviet Socialist Republics,
Washington, D.C., U.S.A.

Dear Mr. Olson:
Your letter of November 17th, addressed to Mr. B.N. Krylov has been sent to the appropriate Soviet organizations for their consideration.

Sincerely yours,
Anatoli M. Goryachev
Second Secretary

This was in reply to a long letter I had sent earlier. The date was a bit odd in light of the following letter from Krylov, also dated December 16th. On December 12th, I had written him a second time, having heard nothing, asking for an appointment.

December 16th, 1958
Embassy of the U.S.S.R.
Washington, D.C., U.S.A.

Dear Mr. Olson:
This the reply to your letter of December 12th. I intend to be in Washington on the dates you mention in your letter and will be delighted to have a meeting with you.
Would you be so kind to telephone me upon your arrival to Washington to fix the exact time of our meeting?

Very sincerely yours,
B. Krylov

This was followed by the cordial meeting at the Embassy which was mentioned earlier, and as a result of which I wrote to Professor Orlov.

Chicago,
December 30, 1958

Dear Mr. Krylov:
I am enclosing a letter which I hope contains all of the information that you may need concerning my prospective trip for study in Russia.

You will note that there are three reprints of papers enclosed. It occurred to me that these might, better than anything else, show the nature of the work I am doing.

I would like again to express my appreciation of the interview that we had on the 27th of this month. I will, of course, await with interest developments as they pertain to this trip.

> Very sincerely yours,
> Everett C. Olson

> January 19, 1959
> Embassy of the U.S.S.R.
> Washington, D.C. 1959

Dear Mr. Olson:

This is to acknowledge the receipt of your letter of December 1958.

We have forwarded it to the appropriate Soviet organizations for their considerations.

> Sincerely yours,
> Anatoli M. Goryachev
> Second Secretary

> Chicago,
> February 18, 1959

Dear Mr. Krylov:

As you will recall, I wrote you early in January 1959 with reference to my proposed trip for scientific study to the Soviet Union. The letter followed our conference of a few days earlier in Washington, D.C.

Since I hope to begin this trip approximately May 15th, 1959, I shall need to make reservations for transportation and also to take care of other matters pertinent to the trip. Thus I am writing to inquire if there is any information available with reference to the request for clearance of this trip with the Soviet Government.

I will sincerely appreciate hearing from you on this matter.

> Very sincerely yours,
> Everett C. Olson

Some time after this, getting rather worried, I telephoned the Soviet Embassy, only to be told that Secretary Krylov was no longer there.

March 2, 1959
Embassy of the U.S.S.R.
Washington, D.C.

Dear Mr. Olson:

This is to acknowledge the receipt of your letter of February 18, 1959.

We have forwarded your request to the appropriate Soviet organizations for their consideration. We expect an answer at any moment.

Necessary arrangements for the trip to the U.S.S.R. can be done through an American Tourist agency dealing with the Soviet Tourist Agency "Intourist."

Sincerely yours,
Anatoli M. Goryachev
Second Secretary

This was a refreshing addition. I never found out what these organizations were or even if they existed. Whether my efforts had any effect, or were at all necessary, has never been clear. I did, of course, get a visa, as suggested, without a bit of trouble. I have a suspicion had I just gone to an American tourist agency and arranged everything through them, that things would have been much the same. At least on all later trips, except for one official one, this is what I did. Only on the official trip did I have a snarl on visa and passport, becoming a nonperson for several days in Moscow, both to the Soviet authorities and our embassy.

When in Moscow on my first visit, I was given a lot of freedom, able to dismiss my Intourist guide so she would not sit for eight hours a day in the Museum, be able to leave the city for the summer dachas of the Academy of Science, and so forth. While I was away, Professor Vistelius, of Leningrad, came to the Hotel Ukraine to find me. He was amazed, he later told me, to find that Intourist had no idea where I was.

Correspondence with Efremov

Efremov (Figure 20) and I had corresponded at length prior to my initial visit to the USSR in 1959. This early correspondence carries the flavor of our mutual interest in science. Only later did our letters shift into other areas. We began by discussing, via the mails, the similarities of the Russian and American

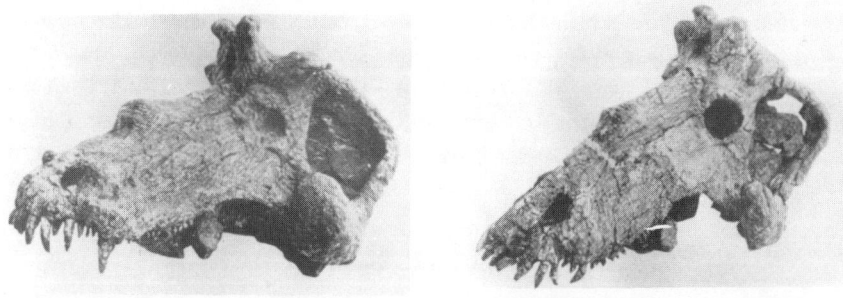

Figure 20. Top: Formal portrait of the "heroic" Efremov. **Bottom:** Juxtaposed, two skulls of *Estemennosuchus uralensis*, a "giant" dinocephalian from the Ocher site. These reptile skulls are about two feet long.

Permian. Some of this comes out in the following excerpts from letters written between 1956 and 1959. Here, and elsewhere, I have quoted verbatim from our letters, with only minor editing of Efremov's for clarity, for his free-swinging use of the English language was a charming characteristic of the man.

August 1, 1956
Chicago, Illinois

Dear Professor Efremov:
    Your letter and the accompanying materials arrived safely. I sincerely appreciate your willingness for me to publish summaries and translations of your papers in English. I have nearly completed a summarization of your "Catalogue" and will submit it to the Journal of Geology. I am reducing much of the detail on specific localities and using maps for their locations. I will, of course, give you full credit for the work, but will assume full responsibility for the translation.
    I am very grateful for the manuscript that you enclosed. The finds of a captorhinomorph and caseid are extremely interesting. I will await with interest your full description of these materials and hope that other groups that tie to North America will show up. It is gratifying to find that at least our sections do have some degree of overlap.
    This spring, as I may have noted earlier, we opened up a new quarry in the San Angelo. At present we have found nine genera, at least four of which are new. They include a Labidosaurikos-like form and three rather small, carnivorous pelycosaurs (?). The excavations are far from complete. I hope to finish this quarry next spring. It will add materially to the known San Angelo fauna.

With best regards,
Everett C. Olson

The translation mentioned in the first paragraph was of *Catalogue of the Localities of Permian and Triassic Terrestrial Vertebrates in the Territories of the USSR* (with Dr. B.P. Vjuschkov). It was, and still is, the only compilation of the Permian vertebrate faunas of the Soviet Union. Also this was my first venture into translation after laying aside Jacques' *Russian Primer.*

The manuscript sent to me reported for the first time the members of the reptilian families Captorhinidae and Caseidae in the Soviet Union. About a year before I had made an analysis, in a volume in memorial to K. P. Schmidt of the Field Museum of Chicago, that noted as important the absence of both of these families in Europe. The timing of Efremov's work was perfect to show immediately that I was wrong. Later, when I knew the

Russian paleontologists well, they would laugh about this. But I told them that we Americans were not at all surprised because we were used to such "communist dirty tricks." By then they thought this was funny.

> Moscow, October 8, 1957
>
> Dear Professor Olson:
>
> I have just received a single copy of your translation of the "Catalogue." You have bite off a big chunck with excellent results. You have very satisfactorily summarized all important data and cleared misty points. The translation is exact, as well, and also the transcription in Cyrillic, except in the list of literature. Here is a rare photograph concerning the taphonomic process: a herd of hippopotami doomed to perish in a small ooze pit amidst a vast, muddy plain drying up after an overflow.
>
> It seems to me to be a typical example of embedding of the large, Permian, semiaquatic reptiles. I hope this year you succeed in discovering true therapsids in the San Angelo.
>
> Sincerely yours,
> I.A. Efremov

I breathed a sigh of relief at the reception of my English version of the *Catalogue* . . . . It had been sent to many places around the world and was beginning to engender a new interest in the Russian Permo-Triassic. It would have been a minor disaster if Professor Efremov had not liked what I did, or had found serious errors in my summations.

The term *taphonomy*, briefly mentioned in the preceding letter, was coined by Efremov in a short article published in English in 1940. The article, which had gone by largely unnoticed, was followed by a book on the subject in 1950, but this publication was in Russian and few in the US paid any attention to it. A fine French translation was made, but it, too, received little notice. Taphonomy is roughly the science of burial and refers to the accumulation of fossil deposits, the transition from living populations to burial sites. It is self-evident that the remains of organisms buried at a site include only a minor portion of the living population from which they were derived. The big problem of reconstruction of the living system from the dead and buried depends on taphonomic analysis, on unravelling the processes of burial as clues to the ways that transportation and deposition have introduced biases into the samples.

It is fascinating, as a practical look at a taphonomic process, to stand near a raging torrent during flood times and see what rushes by—trees, bushes, mud, stones, boulders, a floating carcass of a steer, drowned birds, tumbling clams, or perhaps a struggling snake or rat. Where did they come from? Were they all living together? Where will they end up? These are taphonomic questions. The questions existed long before Efremov coined the term taphonomy but, as is so often true, it was only upon definition of the problem, and use of a catchy term, that attention focused on the subject.

Slowly, during the 1960s and 1970s, after being pushed by some of us—both vertebrate and invertebrate paleoecologists—taphonomy has become a subscience in its own right. The two main areas of interest in taphonomy, the life zone and the burial zone, are in a sense opposites. They appealed in this context to Efremov's own unique sense of a dialectic. In none of the writings that he did on the subject does this dialectic of the source and depositional areas come out clearly, but this was the foundation on which the concept of taphonomy was based.

Chicago, October 16, 1957

Dear Professor Efremov:

Thank you very much for your letter concerning the summary of the "Catalogue." I am most pleased that you found it moderately satisfactory. I am sending you 10 copies. I have many requests from several universities in this country, from England, France, Finland, Germany, South Africa, Australia, and so forth. As you can see, there is widespread interest in your work but apparently too little ability to read your language.

May I thank you for sending me your book entitled "The Land of Foam." I have just finished reading "Doroga Vyetrov" (The Trail of the Winds). Your non-technical Russian poses some difficult vocabulary problems to me. My book with Robert L. Miller, entitled "Morphological Integration" is due off the press next month. I hope soon to reciprocate for the books you have sent me by sending a copy. It is not the enjoyable sort of book like "The Land of Foam" but perhaps some of the ideas it contains will prove stimulating.

Please give my regards to Professor Orlov and tell him I will be writing him soon.

Very sincerely yours,
Everett C. Olson

Moscow, February 25, 1958

Dear Professor Olson:

I am extremely interested in your newest work entitled "Morphological Integration." At any rate, we need a single copy for our whole group, including Professors Orlov and Obruchev.

In answer to your question about "Laurasia" as I have used it, it is an old European term for a great northern continent counterbalancing to Gondwana and including Laurentia and Eurasia (Erie + Angarida + Sinia). Such a continent, however, never existed, as the Gondwana. And you are extremely right to say that Gondwanian faunal elements were embedded in your Permian facies, distributed somewhere on the northern continent. My paper is " Гондванские Фации Северных Материков " (Gondwanian Facies of the Northern Continents). It is a pity that you have not read my book "Тафономия и Геологическая Летописъ" (Taphonomy and the Geological Record) that I sent some years ago. You are possibly the only VP who stand very near to all taphonomic ideas and can clearly imagine all significance of this regularities.

Yours sincerely,
I.A. Efremov

I actually had read a good bit of the book during my early days of learning to read Russian. I probably missed some points and it must have shown in my letters. It does bring up a point which is always bothersome. How much of what I took in and absorbed as my own actually came from other sources not acknowledged, even to myself? This is especially true with respect to graduate students, for mine, at least, have been a constant source of ideas and stimulation. I think that all who work creatively must face this problem. There is, of course, deliberate lifting, but this is rare and most of the "lifting" is not planned or recognized. In a later work—*Vertebrate Paleozoology*—I credited Professor Efremov as one of the major influences in my work from the 1950s on. Some of my colleagues wondered if this was not overdrawn—mainly, I think, because they have seen in me a Russophilic tendency, possibly with some justification. But such letters as this one and many others attest to the reality of Efremov's strong influence on my thinking.

Moscow, April 26, 1958

Dear Professor Olson:

Many thanks from my colleagues and me for your interesting papers. The last paper of the Vale-Choza is real taphonomy. Your other great book "Morphological Integration" is somewhat above my understanding. It seems to me that such extrapolations are steps to the future science as well the heredity of cybernetic mechanisms—mesons and so on. I will give the book to my more competent colleagues.

We are now planning a big paleontological expedition to Central Asia together with Chinese vertebrate paleontologists, beginning in 1959. But personally I shall stay in Moscow because of my heart disease.

As far as visiting Permian localities in this country, it will be difficult to arrange as all of our field staff will be absent in Central Asia.

With best wishes,
yours very sincerely,
I.A. Efremov

Morphological Integration dealt with an attempt to understand the integration of the vertebrate body relative to its functions and evolution. It was done strictly on the basis of measurements and statistical analysis. A mathematical model was developed. The math seems somewhat primitive and crude today, but was, to us at the time, sophisticated. Computers with the capacities needed for clustering did not exist when the book was written, and computer science had not yet found ways to program our data in ways we desired. We did it by hand! Efremov, like many of our colleagues, did not understand the book. Later, crystallizing his historical, naturalistic, geological outlook on paleontology, he decided that this book and such works provided misleading bypaths and had no future in understanding the "real" materialistic world. The vogue of modelling today refutes his feelings as to the persistence of this kind of study but, of course, the jury is still out on its ultimate consequences.

My study visits to the Soviet Union, from 1959 to 1971, came after this correspondence. During these visits I came to know many of the Soviet paleontologists well, and to establish close and lasting friendships. However, it was only with Professor Efremov that I developed a rapport that carried well beyond science.

Correspondence with Orlov

Chicago,
February 21, 1957

Dear Professor Orlov:

Sometime back I wrote Professor Efremov about our general plans. As Professor Romer wrote earlier, he and I are hoping to visit your country during 1959. I mentioned to Professor Efremov that I hoped it would be possible to see some of your Permian localities, while in Russia. I have not heard from him on this matter. Perhaps he has been away, but I would like to know about the possibilities.

Very sincerely yours,
E. C. Olson

Chicago,
September 30, 1958

Dear Professor Orlov:

You will recall that last winter I wrote to you indicating that I hoped to be able to come to the Paleontological Institute sometime during 1959. I hoped Professor Romer would also be taking the trip, but as I believe he told you in London, he finds it impossible. My plans have come along very favorably, for I have recently received a grant from the National Science Foundation[1], supporting my Permian studies and including funds for study in Europe, including the Soviet Union.

I assume this is still satisfactory to you, as indicated in your kind letter of some months ago. It would probably be of considerable help at this end, in arranging a visa and so forth, if I had a somewhat more definitive invitation to come to Moscow and study your collections. I don't know, of course, what your policies on such matters may be, but anything possible along these lines would be helpful.

Very Sincerely yours,
Everett C. Olson

---

[1] I want once more to express my gratitude to the National Science Foundation for aid in this project and many others related to it. I wrote the proposal for a grant in a wet tent in Texas during a rainy spell on Ignorant Ridge. I sent in a handwritten copy to my invaluable secretary Odessa and she did the rest. My handwriting is notoriously bad. The NSF ignored such oddities as "mimeographic" for "monographic" and a budget which I pulled out of the Texas air and, with appropriate monetary modifications, approved the proposal. This was a time when the NSF was fairly new and had not yet become seriously embroiled in the escalating bureaucracy.

Moscow,
November 27, 1958

Dear Professor Olson:

I have received after my return from China your letters dated September 30 and November 3, 1958. Thank you very much for the letters and your fine words about my work on the Titanosuchia. Now about your supposed visit to Moscow. It seems to me your arrival ought to take place at the time of Efremov's stay in Moscow and if so—the only time in 1959 is April, May, June. No sooner because of Efremov's supposed leave during January-March and later also. Can you visit Moscow in the period of April-June.

What concerns some official invitation from the Academy of Sciences I cannot, very sorry, to promise to arrange this invitation because according to our rules and "modus" it means <u>payment</u> by the Academy of Science of the USSR.

Indeed the Academy is restricted in sums for invitations and the very same time a very long "queue" or line exists on the desks of our officers because many scientists from many countries would like to visit Moscow and the USSR in general. Very probably the United States-Soviet agreement of Cultural Exchange will be of use to arrange the general situation. But I do not know in what manner it ought to be done from our side.

Very Sincerely yours,
Yuri A. Orlov
Director, Paleontological Institute

Chicago,
December 12, 1958

Dear Professor Orlov:

Thank you very much for your long letter which I received recently. I am looking forward to receiving your paper on titanosuchids. Very fortunately I had a copy of the Memoir you need in my library and I have sent it to you under separate cover.

I understand completely concerning the matter of an official invitation from the Academy of Sciences. I had not realized the implications with respect to funds. As I noted I have sufficient funds through the National Science Foundation and I am sure your personal invitation will serve my needs. I hope to visit your Embassy in Washington later in the month to make arrangements for my trip.

The months you specify from April through June fit my schedule well. I certainly do want to see and talk with Professor Efremov. My main interest, of course, is to study your Permian materials from the lower part of your section, but I am interested in all of your collections.

Again, I thank you for your kind reply to my letters and apologize for the trouble I must be causing you on this matter.

Sincerely yours,
Everett C. Olson

Moscow, January 15, 1959

Dear Dr. Olson:

Thank you for your letter of December 30, 1958. I have received yesterday. I am glad to learn about your conference with Secretary Krylov. Do not think about troubles your visit could mean to me or Efremov, we shall be very pleased to see you in Moscow. I cannot organize any invitation, but as I explained in my last letter, it depends not on my or our administration's conception, but on the economy and sums devoted to invitations from abroad.

What concerns reciprocal visits it can be of mutual use to our interests, but in as much as I know—it is rather difficult problem as a result of very expensive life in your country (and) restrictions for Russians to visit in your country.

Cordially,
Yuri A. Orlov
Director, Paleontological Institute

Chicago, May 4, 1959

Dear Professor Orlov:

Thank you for your recent letter. I certainly appreciate your suggestion that you meet me at the airport. I would not like to put you to this trouble, but on the other hand would not want you to feel that I would not appreciate your kindness in meeting me.

I have my tickets and visa and will, of course, be under the auspices of Intourist in your country. Hotel reservations have been made but I don't know at what hotel I will be staying.

My schedule calls for arrival at 10:10 P.M., Sunday, May 17th, by KLM tourist flight. This is quite a late hour and I should not like to inconvenience you by it. I may be some little time at the airport since there will be matters of papers and customs. If it is quite late when I arrive at the hotel, I will get in touch with you in the morning.

Sincerely yours,
Everett C. Olson

Figure 21: At the old Paleontological Museum, Moscow, 1959. **Top:** Academician Yuri A. Orlov, director of the Paleontological Institute, I, and Professor Konstantine Flerov—director of the Paleontological Museum, in the background—in a jovial mood during intermission at a doctoral exam. **Bottom left:** Flerov and I listening seriously to the defense of the dissertation. **Bottom right:** Portrait of Professor Orlov, from his biography.

As implied in this last letter, I was superimposing my ideas of scientists coming to this country with what transpires when the process wanes in the other direction. We usually find out by personal exchanges just when our visitor is to arrive and, unless he lets us know, we find out as a rule when he makes a phone call. The government is not involved unless the exchange is a formal one. So I had assumed that Professor Orlov (Figure 21) might be in the dark as to my exact plans, as much as I was. This, of course, was all wrong. Never in the several times I visited the Soviet Union was my agenda not fully known to my colleagues there. Going as I did, by the simple way of being a tourist, I was always under the aegis of Intourist, and everyone but I knew all about what was going on.

8

Impressions — Efremov, Museum, Friends

The first time I saw Professor Efremov he was standing in the hall of the Paleontological Museum partly obscuring an absurdly horned, grotesque skull of a Permian reptile named *Estemmenosuchus*. He was a big man, robust and dwarfing the ancient giant. The two, each unique in its own way, have incongruously stayed linked in my memory. Now both are gone into obscurity, the one some 250 million years ago, the other in 1972 in a country where his name may be enshrined or erased by political whim. Russia and the world are poorer for the passing of Efremov. I really see no one standing on his shoulders to think boldly and imaginatively of better times to come and—in thinking—making them happen. But the successors of *Estemmenosuchus* rose to heights, and perhaps some Phoenix will arise similarly from the dim contrary of the dialectic of human knowledge so dear to Professor Efremov's heart and mind.

Scientist, philosopher, writer and Russian to the core, Professor Ivan Antonovich Efremov seemed to me a unique phenomenon in a society where such a free, imaginative spirit had no business to be.

Like many Russians, Efremov was frank, honest, disconcertingly direct and meant just exactly what he said. Russian, translated directly into English, often comes through harshly, except in the hands of a master translator. When Efremov put the lovely phrase "sadyetis pazhaluista" ("would you please be seated") into his English, it came out "sit down" or "take your place." One sat! When he felt my book *Vertebrate Paleozoology* was too biological and not enough geological he wrote that I was a "traitor" to my profession. Biological analogies and abstractions were to his strictly materialistic mind idealistic fantasies, anathemas to the ingrained doctrines. Oddly, in his writings, he used them. As our friendship ripened, many contradictory ideas emerged and some of his basic ideas were quite alien to me. We both enjoyed trying to sort them out, to see if they could be reconciled. It was refreshing that Efremov's thoughts were put forward in his blunt way and not couched in the confusing jargon of ideologies. We wondered if our concepts, his deeply rooted in an odd dialectic and mine in the more linear mode of western science, were really so different. In long conversations and letters, I managed to learn a lot about him, his ideas and the environment in which his ideas developed. The substance of these conversations, often flavored by a "few drops" of cognac, and his writing about ideas, under my prompting, provides the substance of the rest of this book.

Professor Efremov and I had come from very different backgrounds and I am sure that he, as I, was curious and perhaps somewhat apprehensive about what might emerge as we first met. We had, of course, come to know each other's scientific ideas in exchange of letters and papers, but even in science backgrounds and personalities carry a heavy burden in development of mutual trust and cooperation. From the time of our first meeting in the Paleontological Museum in Moscow, a strong sense of rapport developed. We both spoke of this many times later, curious about just what was involved. Whatever it may have been, as far as I was concerned, his influences on my outlooks and researches in subsequent decades has been profound. Nothing, of course, develops in a social vacuum and it was in the atmosphere of the Museum, the kindnesses of the scientists there, and the complete freedom to study the collections, that our friendship matured and explorations of each other's ideas became possible. A little background, in the sense of "my time in Moscow" during the first visit, will set the stage.

The Museum Scientists

When I first visited Moscow, the Paleontological Museum was housed in the left wing of the old Mansion of Orlov dating from the time of Catherine II. The main building was occupied by the Presidium of the Academy of Sciences of the USSR and the coachhouse wing housed the Paleontological Museum at one end and the Mineralogical Museum at the other. There is a new museum building now, but I miss the old one. It was in the midst of its clutter of bones and display cases that Efremov and *Estemmenosuchus* were juxtaposed during one of my early days studying there (Figure 20). Even at this time, Efremov had a bad heart and was spending much of his time, to the dismay of Professor Orlov, in a dacha of the Academy of Science writing novels.

Just inside the front entrance to the Museum, a giant skeleton of a duck-billed dinosaur from the Gobi Desert formed something of an arch over the door of the office of the gracious Director of the Museum, Konstantine Flerov. Beyond in the main hall, a near jungle of cases of skeletons and bones, many from the Cis-Uralian areas and some from the Gobi Desert, rose in ordered disarray. This was what I had come to study. In one case, some 35 skulls of a primitive pareiasaurian reptile were stacked in a pyramidal fashion.

In this museum were the treasures of the Russian Permian which, except for the most general facts about them, had essentially been "lost" to the western world until about 1954. Since then, many other scientists have passed through the doors during their own visits (Figure 22). The main barrier, language, still remains, and few publications other than the technical *Paleontological Journal* are regularly translated into English. Year by year circumstances improve, but much of the volume and detail of Soviet day-to-day science remains obscure.

Soon after I had arrived at the Museum, most members of the staff departed for Peking, China. Professor Yuri Orlov, after meeting me at the airport, took care of all my problems and became a close friend (Figure 21). Dr. Peter Chudinov (Figure 23), who also studies Permian vertebrates, was among those who left for China, but during later years he became my close companion and guide. We became fast friends. He is a medium sized man, slender and strong, and has a "Uralian" cast to his appearance, revealing his ethnic heritage. His manner is closely

Figure 22. Top: Field camp at the Isheevo site, source of some of the finest Permian vertebrates ever collected, in 1935. Yuri A. Orlov is seated at left. **Bottom:** At the entrance to the old Paleontological Museum on the grounds of the Presidium of the Soviet Academy of Sciences, 16 Leninski Prospect, in 1970. **Left to right:** P. K. Chudinov, L. P. Tatarinov, A. A. Viripichnikov, B. A. Trofimov, K. K. Flerov., A. S. Romer (of Harvard University, who was visiting) and D. V. Obruchev.

Figure 23. Peter Chudinov and skull of large titanosuchid reptile, a "sabertoothed" carnivore, the posterior parts of which had been reconstructed. Photograph taken in the new Paleontological Museum on the outskirts of Moscow.

self-contained and his smile slow, guided by a subtle sense of humor. His generosity in time and hospitality seemed endless. During my many visits to the Soviet Union, I came to know his wife and two daughters well, and watched the children grow to maturity through the 1960s and early 1970s (Figure 24). Peter and I still are close, but now mainly by correspondence. He was Efremov's protege and had nearly completed his major work for the Doctor's Degree (more or less equivalent to the Doctor of Science Degree in Western Europe) at the time of Efremov's death. He was left without a sponsor, and I, at his request, tried to fill in a bit by way of writing a long letter to the Institute about his work, particularly about some controversial points. How much my effort may have helped I don't know, but he did pass the rough hurdles and receive the degree.

Dr. V. G. Sukhanov, who has made a name for himself in functional anatomy, was a young scientist when I first was at the Museum. He was assigned to me to handle the chore of taking care of my needs for specimens, literature, papers, writing materials and so on. We, too, became friends, but somehow

Figure 24. A feast at the Chudinov's apartment on Vavilov Street, Moscow. **Top:** Professor Efremov, in a jovial mood, and Inna Chudinova, Peter's wife. **Center:** Professor Orlov receiving some deep points of wisdom(?) from me. **Bottom:** (left to right) Peter, Yuri Orlov, I and Efremov around the bountiful table of fruit, wine and cognac.

his rapid Russian escaped me and my English was no help to his second language, which was French. We had some long and interesting, if futile, pauses where language and pantomime all break down and the only thing to do is to say "forget it" and either walk away or try to start over. Once things get off in this direction, whether ordering wine and cheese in Greece in unfamiliar French, or tortillas and beer in Mexico, there is no momentary cure. But without Sukhanov's searches for specimens and the literature about them, my results would have been cut in half.

While rummaging through collections from a choice Permian locality I came across some "lovely" coprolites, that is, fossil excrement or faeces. The interesting thing, beyond the rare occurrence of undigested bones and teeth, was that part of the diet of the amphibians from which they came included remains of their brothers or sisters. They were cannibals! I wanted to tell someone about it.

My Russian included a knowledge of good, solid expletives learned from Matthew Nitecki, of Polish origin, who learned them from his father who did not want his children to learn vulgar Polish. These got mixed up with the coprolites as Dr. Sukhanov and I discussed the day's work on the Museum floor. I told him excitedly about the coprolites.

"Coprolites?" he asked? "Shto oni?" ("What are they?")

"Coprolites," I repeated, louder.

"Ne ponimayu 'coprolites'," he came back. At this cross-tongue impass, I said loudly, as if it might help, "Iskopayamoe govno," which in Russian is the equivalent of "fossil shit." My Polish assistant, Matt, would have been proud of me. Not my Russian aide.

"Ya panimayu," he said, glancing around, red-faced and dragging me away from people who were staring.

"Coproleet!" he explained outside, for there is no "-ite" in Russian and it completely threw him.

Dr. L. P. Tatarinov, now an Academician and Director of the Paleontological Institute, is a keen, imaginative student of fossil reptiles and anatomy. The late Professor P. Obruchev, the "English-speaking" member of the staff when I was at the Museum, worked on the extensive collections of ancient fossil fishes. At the other end of the biological time scale was Professor Flerov, whose main area of study was Late Cenozoic mammals. Between these two extremes was the work of the dinosaur expert

Dr. A. K. Rozhdestvenski. During one of our visits, my wife, Lila, spent two hours in a conducted tour of the dinosaurs with him. Both seemed to enjoy it immensely, although how they communicated without a common language I never knew. Lila can do this with faith that somehow her English will be understood, and somehow it seems to be. She is successful in all languages except French, which she knows to some degree.

Dr. M. A. Shishkin is a brilliant, hard working paleontologist who studies Triassic amphibians. With him, Dr. Sukhanov and Dr. V. A. Trofimov, I went through a microscopic display of newly acquired Cretaceous mammals from the Gobi. The materials are remarkable, but each great discovery under the scope was greeted with a typical Russian toast in cognac. The later ones looked even better than the first. Dr. A. G. Sharov was a young student of paleontology at the Museum when I was first there. He later became famous for his description of a "furry" flying reptile and a weirdly scaled lizard-like animal from Kazakhstan.

Dr., or Madam, Konzhukova, Efremov's wife in 1959, before her sad death, was an expert on the Permian and a most gracious hostess both in the Museum and at home. For the first two weeks in the Museum, with white linen and silver and Madam Konzhukova as hostess, we had tea at about 3:30 each afternoon. Cognac, of course, was optional. I rather wondered how much work went on after tea. At first this seemed the British style tea, with a long work period following, but as far as I could see, tea was rather the end of the work day at the museum. The work day seemed to start at about 10:00 A.M. I did keep on working until about 5:30 and someone always stayed to see me to my Intourist cab. I considered this a real courtesy, and still feel so. My more suspicious friends assume it was surveillance. Possible, I suppose, but I prefer to approach the world more simply, abhoring unnecessary and nonproductive suspicion. One gets burned and duped now and then along the way, but life's much more pleasant without the unnecessary paranoia.

Evenings in the Hotel Ukraine on the Moscow River were long and dull, mainly good for reading and trying to comprehend the radio piped into the room. The vast lobby with the big Intourist salon to one side and the massive restaurant to the other was the crossroads of the world at this time. The Hotel was one of the principal Intourist repositories then, with the incessant multilingual babble, a pervasive smell of sausages, and late in the evenings, the restaurant, with the band playing such favorites as *Sweet Sue*, it was the tourists' Moscow. Eating, beyond deciphering the endless menu, was a problem to a neophyte. If one sat at some of the tables, nothing happened. These tables, unmarked, were not being served and one just sat. After learning this I was able to rescue a few stranded Americans from this apparently odd trial by silence.

After two weeks, the teas at the Museum stopped, just when I had begun to wonder if the staff had not found a pleasant new way of life. I have been to many places, many museums, and have, as a rule, been welcomed, but the grace and cordiality of the welcome and treatment in the Paleontological Museum at Moscow has never been approached elsewhere. The teas did wonders in introducing me to members of the staff and to many little Russian ways, even though the repartee left my "schoolboy" Russian far behind. One thing I did find out was that humor doesn't easily cross language barriers.

Someone in the course of conversation had asked me if we really did have instant coffee. Being reasonably egotistical about America I answered,

"Oh yes, we also have instant tea and milk."

"How do you make it?" came the eager response.

"Simple," I said. "Just add water." But then I got carried away. "We also have instant water," I quipped.

"Shto?" burst out.

"Da," I replied, "Good for field work, when it's dry you just add water."

Well, that is not very funny to start with, but it was supposed to be. Only I knew it.

"We don't understand."

"Please explain."

The more I tried to say, "It is just a joke," the worse it got. My flat "American" pronunciation of "water" didn't help. I was not articulating, as my British friend Doris Kermack always told me. I came to feel more and more like an English friend who, soon after World War II and while sitting at my American table, told a war story which involved the punning of "WC", confusing war correspondent and water closet.

"Never," he said glumly after viewing the vacuous stares, "never have I sat through such a sticky silence."

Professor Efremov came to the Museum only occasionally, to see me. So I was given his office, or cabinet, to use for study. This was more or less on the second floor of the Museum, for there were offices on an intermediate lower level. The cabinet was a few steps down from a large rotunda covered by a glass dome skylight. I like to imagine that in the old days the people of this left wing held gay parties with wine, vodka and balalaika music for dancing. In 1950–1960 it was the meeting hall for the Paleontological Institute. Down the hall a ways, and down a few steps, was Professor Orlov's office.

The main building of the Paleontological Institute was down Leninski Prospect about two or three blocks. I rarely saw many of the paleontologists, those whose main operations were down the street. In the Museum, Professor Orlov would drop in from time to time during the day to see if all was well. More often, I would just see him pacing by, muttering his likes or dislikes of what was going on. Dr. Sukhanov would see me early each day, to see what I would need for the day, and drop by now and then to find out if all was well. Daily the custodian would shoo me out of the office, open the windows and clear the air. I smoked heavily then. Whether my attendant thought I would be asphyxiated, or what, I never did find out. When noon came, after the teas had stopped, my odd habit of noon eating was automatically catered to. The usual fare was thick cut French bread, deeply buttered and heaped high with caviar or sturgeon. Lunch was announced by a husky female custodian with a bellowed "kuschits" (pronounced as it looks, and meaning to eat). I never quite got used to it.

In Efremov's cabinet was my favorite picture of him, over my right shoulder, and an autographed picture of Gina Lollabrigida to the side. I tried to get the story of the photograph of the movie star, but never did. One of the tantalizing things about a strange country to me is that one is never just sure what is going on; it all sort of flows by in a muddy swirl with things popping to the surface for a moment here and there. Quick, idiomatic repartee is hard to catch, but in it lies the real, underlying sense of a culture.

Later I found out that Efremov had a sense of beautiful women and a gentle eroticism, which shows up in his non-science writings. At times he asked me to send photographic magazines, with nudes in them for his son—he said. Or, he needed them for artists to illustrate his books. On the surface,

Moscow seemed extremely puritanical and strait-laced. Even the fanciest nightclubs, largely for dignitaries and tourists, reflected this in their 1930s vaudeville entertainment. But this may be all wrong. After a few days or weeks in a foreign country, one seems to know all about it; after many weeks or months and successive visits, it becomes evident that the opposite is actually the case.

This first visit to Moscow was later followed by six more, mainly for study and to help our exchange program. More than anything else, these visits gave me the chance to see and talk at greater length with Professor Efremov. My first visit was made during the freeze of the cold war, during the most crucial of the Berlin crises, when Khruschev threatened to move to expel other powers forcibly by May 27th, 1959, if I recall correctly. It was worrisome. I rerouted my plane reservations to avoid East Berlin, originally my first stop.

But this crisis passed and, when I returned to Moscow later as an old, and I think trusted, friend, the tensions had eased. Yet, I never did feel them directly in Moscow, even when reading *Izvestia* and *Pravda* religiously. Instead, the atmosphere that I encountered in 1959 persisted and deepened, making study and inquiry in Moscow both pleasant and rewarding (Figures 25, 26).

Figure 25. Whether in the US or USSR, we did the same sorts of things—we hunted for bones and had fun. Here, bulldozers cut down to the producing layers in the US (top) and the USSR (bottom).

Figure 26. Field camps in the US and USSR. **Top:** Camp on Texas field trip. **Bottom:** Camp at Ocher (Yeshova) site, in Russian Uralian area, where fossils were prepared at the time they were excavated.

9

Efremov—the Man

In an intricate way Ernest, Wade and Efremov blend together in my thoughts. I am sure this would have pleased the last, because he was a great fan of our mythical west. "Would you kill this Boer thickboned creeper in Texas manner?" he wrote me about a rather boorish paleontologist. Ernest and Wade I can't be sure of. Maybe the connection comes because they all helped me so much in my studies of fossils, but I think it is something more. In meeting and coming to know each I was in a somewhat alien culture, learning new ways, and each was an honest, straightforward if not always gentle, guide through his strange land. Each was an outdoor man, and in his own way knew nature and the land which had raised and kept him—and sometimes, harshly, nearly killed him.

Efremov, while growing up, had advantages that neither Wade nor Ernest enjoyed. He lived in the country in the village of Virtz, near what was then St. Petersburg (now Leningrad), where his father, Anton Kharitonovich Efremov, a high ranking member of the Semenov Regiment, was manager and part owner of a large estate. In our long conversations, and from his writings in introductions to his books, I picked up some of Professor Efremov's feelings about his early life. I could see in these feelings threads that wove into the complex, exhuberant and ever curious man that Efremov remained even while living in the midst of the rather drab, mundane social setting of everyday Moscow during his later years.

The library of Anton Kharitonovich Efremov was extensive, and by the age of six, Ivan Antonovich, with free run of the shelves, was delving into H. G. Wells, Joseph Conrad and Jules Verne, finding a special hero in Captain Nemo and his undersea adventures. Ahab and the great white whale were high in his list of favorites. If a romantic flair is somehow nested in the genes, it was there in Efremov from the very earliest times that he could recall.

Efremov learned to read very early and very well (Figure 27). He became an avid consumer of books and was multilingual. This devotion to reading persisted to his very last days. Friends around the world received requests to send books from long lists which he had compiled but could not obtain in Russia. Many of his correspondents responded so that his apartment was literally lined with an almost unbelievable olio of books in a dozen or so languages, on topics from the ancient history of Egypt through the Ionian and later Greek philosophers. Oriental books were much in evidence. Some of these books were mailed in, but many sent by this route failed to be delivered; others were hand carried to the Soviet Union. Many never made it at all, disappearing into that mysterious void known only to the postal services and the Russian censor system. Orwell's *1984*, Shute's *On the Beach*, Miller's *A Canticle for Leibowitz*, Comfort's *Darwin and the Naked Lady*, Plato, Aristotle, Thomas Aquinas, Conrad, Zane Grey, adventure stories without end, Mickey Mouse, Michener, Al Capp, Allister MacLean and Bertrand Russell were all there. There were no detective stories and no books on Jewish travails.

In 1917, just ten years after his birth, young Ivan Antonovich Efremov saw his childhood begin to crumble with the onset of the revolution, and in the civil war that followed he saw the family break up. As a child, young Efremov was especially close to his father, a lover of nature, and from his father learned always to ask "why?" and to expect and get an answer. His outlook rarely wavered from the positive thought that there were answers to be had. His father was his first great hero, a man of the woods and the land. Later, so Efremov told me, having lost contact with his father when he was very young, he cast himself in the surrogate hero role and throughout his books, the hero tends to be his alter ego.

With the civil war, Efremov's childhood ended, having begun to crumble earlier. Being large in stature, he was able to

Figure 27. Ivan Antonovich Efremov. **Left:** With his beloved books, 1958. **Right:** In the Navy, 1925.

enlist while just 13 years old in the 6th mechanized corps of the army. Sensing that his beloved country was in danger, he felt he must make his contribution. Much of his service was spent on the shores of the Black Sea and the Sea of Azov, where he first saw the life of the seashore and water firsthand. His love of the sea, developed first from stories he had read, grew even more impelling. Dreams of what lies out there beyond the horizons run through all of Efremov's writings, both in science and in fiction. His military career was ended by a near miss of a shell from a British gunboat, which very nearly killed him and left him without speech for some time.

The roots of his steadfast loyalty to his country, his faith in progress, his prejudices and yearnings, and the slightly sad cast of many of his stories stem partly from his badly disrupted early life. The odd schism of his strictly materialistic philosophy, which eschews the use of linear logic and mathematics, and his somewhat mystical constructs of the universe find a base in the dialectical pull of the sea and its mysterious romance and science with its hard core empiricism. He first found a resolution of this schism in science applied to the geological history of the world in a progressive mode, and later in his romantic, often nautical parascientific writings of adventure, mystery and romantic fantasy.

A side of Efremov that somehow rarely appears in his writings, but was much a part of him as a person, is the whimsy such as he showed when explaining to me his rather pronounced stammer. He had just put his arm around my shoulder, the Russians being great huggers compared to us cold Nordics, and said,

"J–j–just b–b–between us gir–gir–girrrls, we b–both better be af–af–af–afraid of the Ch–Ch–Chinese." He was convinced to the day of his death in 1972 that war between the Soviet Union and China was inevitable. Being a bit shy about his stammer, probably aggravated by the sensitive nature of what he had just said, he explained to me that he stammered in English, because the shell that landed near him had come from a British gunboat. Not quite true, of course, for he also stammered in Russian, but a gentle dig to one with a British (mixed with Norwegian) heritage.

Recovered from the shell blast, Efremov returned to Leningrad, and while employed as a lorry driver's mate, he enrolled in a correspondence course in marine navigation. Labor and study, he told me, were the common practice in Russia in the 1920s. At the same time, having read a paper by Academician P. P. Sushkin on the ancient great rivers and fossils of some 200 million or so years ago, exposed along the North Dvina River, he had written to the author. As a consequence, Efremov was given free run of the Museum under the informal tutelage of Sushkin on the lore of fossils.

Efremov's first taste of paleontology was interrupted by his naval apprenticeship service in the Far East and the Caspian. Still uncertain of his career, he returned to Leningrad and his books, entering Leningrad University in the practical areas of geology and mining engineering. The books in the winter and sailing in the summer tugged in opposition.

With Efremov, as is true for so many, a seemingly minor event determined the final direction of his career. Efremov described, in his introduction to his *Stories*, a coincidence of events which channeled him into science and away from the sea. Curiously, one of them occurred at sea.

> For a long time I had hesitated between two professions, the sea and science. Once, during my service on the Caspian, I was returning to Baku by motor boat. The day was unusually serene and sultry. The sea was a sheet of opaque greyish-green glass; a fiery sun hung in the heavens. I was lying in the boat's prow and peering into the depths; the sunlight reached deep into the water, and in some places I could see the bottom at the depth of about 30 feet.
>
> Suddenly I realized that I was looking down at the ruins of an inaccessible submerged town, at walls and towers, which slowly receded under the boat's keel. I already began to make out elusive outlines of streets and houses when the surface of the sea was ruffled by a breath of wind and the vision vanished.

In Baku, on shore, Efremov found waiting a telegram from Academician Sushkin offering him a minor position in the Academy of Science. The die was cast. The lure of the past seen in the undersea images proved too strong to resist. After Sushkin's death, and with his education and several seasons of exploration in back of him, Efremov was asked to take over some of the work of his mentor. With some reluctance he agreed, although feeling inadequate and knowing that museum work would limit his travel and exploration. As the restrictions on travel became stronger in his new position, he began to seek relief in writing about his travels and the peoples he had seen and come to know. But his awkward, colorless prose, according to his own analysis, failed to satisfy him. He could not seem to catch the essence and beauty of nature and of the history he was trying to portray. Reams of paper went into the trash basket.

Scientific writing, on the contrary, flourished during this period. Efremov's first paper came out in 1927, when he was just 20 years old. In 1940 his doctoral thesis, a massive, mature work in the European manner on "A Description of Habits of the Ancient Amphibians," established the paths followed by much of his subsequent research. Many other papers were published between the 1940s and the 1960s, and at the time of his death Efremov was well into the preparation of a popular book on paleontology for the Russian people.

A switch in emphasis began early in the 1940s. Efremov again took up writing stories, this time with more success. Eventually, he came to devote more effort to this work than to science, which led in part to a schism between Efremov and Orlov. This pushed him even further into fiction writing.

A geologist will often run into things that puzzle him or interest him but have to be put on a back burner for a time. Efremov sometimes found grist for his story-mill in these little notes and memories. The basis for his fictional story "Almaznaya Truba" ("Diamond Tubes"), in which the hero discovered rich deposits of diamonds near Yakutz, was one such encounter. Not too long after its publication geologists discovered, quite independently, actual diamond-producing tubes at almost the same place as that specified in the story. I had talked with Efremov briefly about this and gathered that some sort of a story was connected to it.

"Many troubles followed," he laughingly told me.

It was in Leningrad that more of it came out. The people in Leningrad were extremely gracious, but, as always happens, time begins to hang heavy as a guest is entertained. There was a seemingly endless supply of mineralogical museums, which I as a geologist should enjoy. I think I saw them all under the slow, gracious conduct of the directors. One day, as I was being toured about, with full explanations in Russian, the diamonds came up. As usual, after an hour or so, I had tuned out, giving back nods, shakes, "das" and "nyets", a "shto?" now and then, and an occasional "krasivyi" and "ne vozmoshna" ("beautiful" and "impossible").

My guide opened a flat case and pulled out some dark, translucent, unfinished stones. I came to, a bit.

"Diamonds?" I asked.

"Yes," he answered, "from Siberia."

"Professor Efremov's diamonds?" I went on.

Seemingly a little startled, he asked, "What do you know about that?"

"Not much," I followed, with a comment that I had only heard there was an amusing incident.

"Yes, but not too amusing."

He went on to tell me the story as he knew it, one of several versions I eventually heard. Once actual diamonds and diamond tubes had been discovered by geologists, a "Catch 22" situation developed. Efremov was in trouble for his story, first for concealing important economic information from the government, that diamonds existed, and second for revealing in published form, without permission, "state secrets!"

During the 1960s, fictional writing became more and more important to Efremov, and when he resigned from the Institute, for reasons I do not recall, they became a needed source of income. This change in emphasis did not enhance him to some of his colleagues.

"He's too busy writing his novels," Orlov would tell me as an outsider. "No time for science."

"Professor Orlov ought to push paleontology at the Academy more," Efremov would tell me. "He's the only one with an entry and the power, and he sells us short."

Because I was an outsider, both seemed to feel free to unburden their troubles on me and discuss subjects that were more or less taboo in the Institute.

This rift between Efremov and Orlov mended, but only a few weeks before Orlov's sudden death from a heart attack. The two met in a bank, embraced in proper Russian style and seemed once again to be the good friends of old. Efremov was immensely relieved and, I suspect, the same was true for Orlov.

As it had earlier, Efremov's renewed efforts at fictional writing stemmed from enforced inactivity. During an earlier expedition to Central Asia, he had contracted typhoid fever, which had gone undiagnosed by the local doctors. The typhoid would recur, so he told me, about every five years, so severely that he would be hospitalized. It was, I gather, the start of the ill health that plagued all of his later years. While confined in a hospital in Sverdlovsk, in 1942, unable to serve his country in the Great Patriotic War, he began to set down his experiences in story form once again. This time some of his writings were brought to the attention of A. N. Tolstoy, who invited Efremov to visit him, even though Tolstoy was seriously ill at the time. He praised the style as elegant and formal. Inquiring, Tolstoy found the source of Efremov's style in the rigors of scientific reporting, something Efremov had lacked earlier. In most of the translations of Efremov's non-scientific work, however, this style does not come through clearly. Often the stories appear turgid and verbose. The stories as translated, while imaginative and charming, conform too closely to the literal Russian to have the graceful flow that superior English prose could have given them. The translation of *Cors Serpentis (The Heart of the Serpent)* is an exception and comes close to portraying the richness of his style.

Through periods of good health and bad, science and fantasy alternated in importance. Upon the publication of his scientific papers of the 1940s, Efremov assumed a prominent place as an interpreter of ancient earth history, first in Russia and later elsewhere. His first literary efforts, gathered in *Stories of the Unusual*, published in 1944 and translated into English in 1954 as *Stories*, began a successful literary career, if success is to be judged by the popularity of the product rather than the criterion of acceptance by the *literati*.

Out of Efremov's mix of experiences emerged a strange and somewhat enigmatic philosophy, a materialism with a dialectical base and a deep flavor of Eastern philosophies. Somewhere in all of this lay answers to questions that puzzled me, more so as I came to know him better. How could such a person come to be and continue to live comfortably and productively in the controlled society of the Soviet Union? Why have his works been so popular with the people of the Soviet Union and treated so lightly by the literary critics? Where lay the base for the obvious rapport that both he and I sensed at our first meeting and developed so freely thereafter?

10

Efremov, Science and Philosophy

For all of his ventures into fiction, initially as a release and later as a necessity, Efremov considered himself first a scientist and secondly a writer. Only after I had come to know him intimately, between 1959 and 1971, did I come to appreciate the relationships of these two aspects of the man. Many of the paleontologists whom I came to know through these 12 years have now died. Younger men have taken their places, and some of them may not have known Professor Efremov. His student *protégé*, Peter Chudinov, however, has written an in-depth account of the man and his work (Иван Антонович Ефремов) and it has been widely read. Efremov's impact, particularly in the area of taphonomy, has been widely felt in Russia and has been seminal in many other countries. What follows, however, deals with matters that arose in our close personal relationship and not in any heretofore written material. For all of our close friendship and trust, Efremov always called me "Professor Olson," and all of his many letters began "Dear Professor Olson." With Peter Chudinov, on the contrary, it was Peter and Ole. But Professor Efremov was from an older school, and even Ivan Antonovitch seemed too informal to him. Academics must maintain their dignity!

But names were about the only restrictive formality between us, and I began to feel that there was much more than the scientific side of Efremov that might be of as much interest to others as it was to me. By 1970–1971, we had become close friends, and I had received from him copies of many of his scientific and fictional books as well as a large sheath of letters written in his slapdash English. It seemed to me that they might form the basis for something of general interest. I thus broached the idea to him, wanting his judgment and permission. His reply, slightly edited for readability, tells a good bit of how he pictured himself and how his science was at the core of his self image.

Moscow, June 18, 1971

Dear Professor Olson:

About your project with "Efremov's Letters." It seems to me an extremely interesting and good tribute to our V.P. [i.e., Russian vertebrate paleontology] if you can write a book about it because, as with the geological basis of evolution, you are the only man who can afford such a task. But, of course, not under any such title. Efremov as a person hasn't fame enough for his "wisdom" either as a very great V.P. If you can use my letters as characteristic for this or that achievement of our VP or in comparison with opinions of our western colleagues—this has good meaning. But not for the main stem of the whole book! Efremov will note that before his researches, field for instance, U.S.S.R. had five or six localities of lower Tetrapods and after 1950 they numbered over 200, then you will be able to cite one or two letters. If you say that Efremov was first to make a stratigraphical column on Tetrapods, then another look at my letters is possible, and so on. Like that, the letters can be distributed in the general picture of our stratigraphic localities and the peculiar position of our facies—between two great Tetrapod Fauna, Laurasia and Gondwana, and a third province including China! Such a book may have a common interest among all zoological and geological people; if you will be able to encrust it with some personal characterization it can be even more interesting. But "Efremov's Letters" is "not sounded" as our young people say.

As ever, your friend,
Professor Efremov (Old Efraim)

I had not made at all clear what I had in mind. His science and Russian science are important. They are strictly orthodox as far as paleontology and geology are concerned, the same type of work as that carried out in many countries under many philosophies and many kinds of governments. They do not at this point need a lot of elaboration, for thanks to Efremov and many others, the work has been well done and, by 1970, well reported.

What had become important to me was the underlying basis of his scientific thinking, and that of Russians in general and also how this tied into his adventure stories and science fantasy. Some of this, in relationship to science, comes through in his letters, but only rarely in his scientific writings. His social philosophy, romanticism and love of heroes and adventure stand out clearly in his nonscientific writings, but their more intimate bases are to be found in his commentaries and the ramblings of some of his letters. I tried to make that clear to him and, finally, he did sanction the project. I only regret that he did not live to see it underway and to give me what would have been invaluable advice during its preparation.

Forays into Science and Philosophy

The sort of conflict of ideas occasioned by *Morphological Integration* ran through many of the conversations and letters between Professor Efremov and me during the next fifteen years. We dealt, of course, with the straightforward problems of Permian animals and the relationships of the Russian and American faunas, but even these were cast in a broader mold of philosophic differences. I had come from a society steeped in traditions of linear logic and simple confidence in cause and effect; Efremov, from his own complex mix of eastern dialectical philosophy, strict materialism and faith in a social utopia as an ultimate goal. Nicely, these differences did not lead us into ideological stalemates and dogmatic stands, for I was anxious to learn and he, from his dialectical base, could only see the differences as the way to their resolution in a synthesis.

Philosophers have been plagued for centuries about the sufficiency of materialistic science as a pathway to "truth," and occasionally some scientists also have had a go at the problem. As modern science has tended to depart from its tangible and comprehensible empirical base in "material" toward more mathematical, almost "Platonic," constructs of "reality," the troubling questions have either faded away or have become more crucial depending on where one stands. Philosophers from Plato and Aristotle through the Middle Ages and Renaissance to Whitehead, Russell, Hempel, Popper, Reichenbach and many others have been sifting out answers. Some scientists, Einstein for example, have listened and contributed. Others have rejected the "philosophical double talk" and got on with the job.

A colleague of mine at the University of Chicago, the late Bob Miller, started me on a search for meaning when, after some years as a biologist and evolutionist, he abruptly switched over into mathematics and geophysics, going back to an earlier love. It was a shock to me to see him make such a drastic shift, although I had seen his disenchantment with inferential statistics developing. Miller's move disturbed me mostly because it was motivated by his growing feeling that the study of evolution and evolutionary theory had become a sterile, blind alley. This was my main field of endeavor!

My reaction was to plunge into the literature on the bases and structure of evolutionary thought, to see what I might find out. I began to read rather randomly in Popper, Randall, Reichenbach, Russell, Nagel and Hempel, and then was drawn further afield into Whitehead, Dewey, Kant, Hegel, DesCartes, James and so on. This turned out to be a never-ending quest, as one author led to another.

All sorts of questions did seem to ask for answers. What is evolution anyhow? Is it a principle of the universe or a rationalistic concept? Who started the evolutionary concept—Aristotle, Heraclitus, Hegel, or perhaps Darwin as the textbooks sometimes say? Certainly it was not the Romans, not Augustine, not Aquinas. Was the idea alive during the Middle Ages or the Renaissance? I finally ended up teaching a college course in "The History of Evolutionary Thought," trying to get things straight in my mind and, hopefully, interesting and educating a few students. John Campbell, Clifford Brunk and I explored these concepts in greater depth in a graduate course at UCLA during 1987–1989. The rapid accumulation of data, new concepts in genetics and molecular biology, cladistic analyses applied to systematics and phylogeny and changing ideas in evolution continue to present interesting and frustrating problems. The concepts basic to my exchanges with Efremov were, of course, those of the 1960s, but the philosophic problems that were bothering us are still as significant today as they were then. Mainly some of the questions have changed.

During the time when these matters had begun to bother me, I was going on with paleontology as if nothing had happened. I was also becoming more engrossed with Russian paleontology. My friendship with Professor Efremov had reached a stage where we had high mutual trust and could communicate freely. The interest in the paleontology of the USSR and readings in its literature sent my binge of philosophical inquiry off on a tangent, into the sphere of dialectical materialism. This bypath, and Efremov's thoughts on it, began to take on primary importance to me.

All of us in the US during the late 1950s and early 1960s knew there was something called dialectical materialism and had some vague recall of Socratic dialectics from some dimly remembered college course. Somewhere along the line Karl Marx had "stood Hegel on his head." The Marx-Engels-Lenin philosophy had become the cornerstone of the Soviet Union's socioeconomic system and had run rampant through its science, and that was bad.

The near death of Russian genetics, we had heard, was due to Lysenko's application of dialectical materialism to agriculture. Dialectical mysteries surfaced in the US from time to time in talks and seminars where the audience was treated to a diatribe of Marxian jargon spouted with consummate confusive skill by some biologically centered red neophyte. Scholars of Russian affairs, of course, were more sophisticated than we and knew better. All of this, however, was more or less how I conceived things in Soviet philosophy and as at least some of my colleagues also felt them to be.

When I traded my Russian primer for Russian paleontological literature, I fully expected to find the dialectic blatantly displayed. My failure to do so puzzled me. Was all of this a hoax? Many of my associates, some very knowledgeable, were sure the little bows to Marx or Engels, or to the great Lenin, in introductions to papers and texts, were just that and no more. They were survival gestures in a somewhat hostile philosophical environment. Maybe so, and sometimes certainly so.

The matter was taken seriously by a number of biologists, especially those of Russian origin or those who had been closely associated with Russian genetics. When Loren Graham, physicist and historian of science, published his book *Science and Philosophy in the Soviet Union,* he was taken sharply to task by the late Theodosius Dobzhansky, one of the outstanding American geneticists of our time. Dobzhansky sharply countered the thesis that dialectical philosophy had a real role in Soviet science. The late Michael Lerner, also with a deep interest and knowledge of Soviet science, took a milder tack, but discounted much that was said, while saying it should be said. Dobzhansky likewise took me to task, over a drink, for my somewhat more favorable review of Graham's book. Underneath this lay an inevitable mixing of science and politics. To even suggest that the crude agriculturist Lysenko was a scientist, which he was not, was to support a position that augmented the strength of a detestable regime.

For all the opinions that bowing to dialectical philosophy was just a gesture, I could not dismiss easily the lack of evidence of the dialectic in paleontological and evolutionary articles. I still had the feeling it should be there. It was not clear to me at the time that a basic philosophical concept must be apprehended in order to be recognized as a substrate in those writings that do not make a point of bringing it out.

Professor Efremov, as I had found out when I came to know him and to read his works more thoroughly, was a firm advocate of a broadly based, largely non-Marxist, dialectical method of arriving at the truth and a strict dialectical materialist in the hard sense of the second word. Things are real and as we sense them. They give the answers to the truth and are the truth, but must be understood in a resolution of opposites. When he wrote scientific treatises his themes and explanations were strongly rooted in a materialistic dialectic, but this was recognizable in his science only if one knew the background. His best known work on taphonomy originated directly from dialectical considerations, but mere reading of most of his writing on the subject does not necessarily show this to be the case.

I had developed a feeling in the late 1950s that I just didn't understand dialectics at all as I read more and more books and articles on Russian evolution and biology. This led me into a long series of sporadic studies to see if I could develop a more sophisticated idea of what dialectical logic was all about. This search has continued to the present, but the significance of dialectics has not been resolved in my thinking as I write in 1989. One part of this effort is documented in Chapter 11 in which I explore the work of Professor Davitaschvili on evolution. In order that this and references to Efremov's dialectics may be somewhat clearer, the following is a short summary of some of the things I found out.

First of all, a good summary of the relationships of dialectical thought to biology is to be found in the final chapter of *The Dialectical Biologist* by Richard Levins and Richard Lewontin (Harvard University Press, 1985). But I began at a much more naive level, and even Levins' and Lewontin's review proved far from simple to me. The concepts of dialectical materialism as used by Efremov are difficult for anyone trained, as I was, in a generally reductionist approach employing the syllogistic logic of orthodox science. A sense of mysticism seems to creep into dialectical interpretations. The tendency to dispose of dialectical materialism as a "bunch of nonsense" is strong and intrinsically appealing. Today, however, one of the growing tendencies in evolutionary thinking is at least to give a hearing to what dialectical logic may have to offer.

As is true for so much of our knowledge, the roots of the dialectic reach far back into history, even to the Yin and Yang of the Far East. More particularly, our Western heritage lies with the Greek philosophers, especially Plato, Aristotle and Socrates. To each of these, however, the term dialectic had a distinctly different meaning. The dialectic of Hegel, from which the Marxian form of dialectics stems, lay closest to the concept espoused by Socrates, filtered through the development of irreconcilable opposites in Kant's "Critique of Pure Reason." Efremov's dialectic shared this general heritage but was divorced from the socioeconomic framework of Marx, while retaining the strict materialism. The pervasive flavor in all of these related treatments is that the dialectic is an intellectual process by means of which inadequacies of popular conceptions, in our case scientific conceptions, are examined and exposed with closer approaches to the truth emerging. This sounds somewhat like the scientific process of the erection of hypotheses, followed by testing and acceptance, improvement or rejection. It is not, but this same suspicion of identity keeps cropping up in the mind of a scientifically oriented person.

The basis of dialectics actually is very different and lies in contradiction and negation found in the familiar triadic logic of a concept, or thesis, its opposite, or antithesis, and a resolution, or synthesis. The synthesis is a higher truth than either the thesis or antithesis, but in sequence generates its own opposite;

negation ensues, and a new synthesis forms to generate a never-ending dialectical spiral. In letters quoted in later chapters, I asked Professor Efremov about some of the problems that this approach brought to my mind and he did his best to answer. But, as I found out, this simple framework is far from adequate to give a full appreciation of what he was writing about and how it affected his outlook on evolution.

To go a little further on this matter, it can be noted that, according to the proponents of the dialectic, it is reality, not just a law, and it involves an evolutionary process of inevitable progress. This gives it a particular relevance to biology, with its keystone of evolution. It also poses such problems as progress in evolution, an anathema to many evolutionists, teleology and teleonomy, and a holistic versus reductionist approach to evolution. Such problems, of course, arise in evolution as viewed by most scientists, but in dialectics we are in a form different from our broadly Cartesian approaches, or what Efremov calls "linear logic" in his letters. At the very least, dialectics has stimulated some interesting biological points, such as the following.

From the reductionist point of view, the whole is made up of parts with their properties causally related to the whole, but also potentially having independent existences which have significance when separated (or alienated in dialectical terms) from the whole. Thus a molecular biologist may study DNA or a gene as something interesting either in itself or as it bears a causal relationship to some larger unit. The DNA, gene or some grosser parts of an organism can, of course, be arbitrarily removed from the organism and analysed to give appreciation of a particular aspect of its host.

In the dialectical sense, however, such detached elements are not viewed as important as entities in themselves, but only as parts of the whole which lose their meaning when viewed alone. Also important in dialectical evolutionary thinking is the idea that the whole itself exists in a temporary balance of internally opposing forces in which cause and effect are interchangeable. This is crucial to Efremov's thinking in his evolutionary philosophy. Levins and Lewontin in *The Dialectical Biologist* sum it up on page 277 as follows: "The difference between the reductionist and the dialectician is that the former regards constancy as the normal condition, to be proven otherwise, while the latter expects change but accepts apparent constancy."

Many questions arise about possible impacts of dialectical thinking on the ways in which biology and evolution are viewed. Efremov clearly was deeply influenced by the importance of uniting opposites in a temporary synthesis, as in his taphonomic work which examines the biological and physical factors in the formation of fossil deposits. Throughout evolutionary studies, no matter how they are viewed, antagonisms between the organisms and the physico-biological environment are self-evident. So wherein lies the difference?

One place where it seems to emerge is in the concept of adaptations of organisms to their environments. If this is viewed in a nondialectical manner, with cause and effect explicit and not interchangeable, it is a one-way action, with organism adapting to environment. Is this the full picture? Certainly it is central in the Synthetic theory as developed today, with direct roots reaching back to Darwin in 1859. Yet there are those who see Darwin's total explanation as dialectic, in contrast to Lamarckian ideas which have a strong linear cast. The problems of the adequacy of nondialectical adaptationist approach has had an impact in questions now being raised about the sufficiency of the Synthetic theory.

Ecological studies are complex and deeply involved in development of syntheses of interactions of organisms and the physical world in which an interrelationship of the two in a complex of causes and effects stands out. Like so much in biology, disparate elements merge into a complex—a temporarily stable whole—extremely difficult to visualize by analysis of its component parts. Adequate syntheses are elusive. Would a dialectical approach help?

In these, and other similar broad biological problems, dialectical thinking, if nothing more, has, it would seem, performed its primary function of serving to point out possible inadequacies in some prevailing interpretations and provide an alternate way of thinking.

My puzzle, then, in not finding dialectical bases in the Soviet writings as I first read them did stem in part from my own lack of knowledge of the matter and, in some part as well that, as far as science is concerned at an operational level, it makes little difference what a writer's philosophy may be. An atheist, a devout Catholic, Jew, Muslim, a Buddhist or a dialectical materialist can be a good scientist and communicate easily with his disparate counterparts. Just play the game and be sure to follow the rules. Only when one comes to the fringes, where science *per se* can have no answers, do the roots begin to show through. But at a deeper level this may be less true, for, from this fringe area the *"a prioris"* of the philosophical base do feed back into the methodology of inquiry. A Bible-believing scientist, for example, will see time very differently from an agnostic or one who interprets the Scriptures to fit scientific knowledge. But, unless time is pertinent to the work at hand, this will not show up. With respect to strict dialectical materialism versus formal logic—which were the major polarities to Efremov—there exists a seemingly unbridgeable gulf. Our "hard" mathematical sciences, which involve a simple binary yes or no, cannot accommodate the tenets of dialectical materialism. The "soft" naturalistic sciences, evolutionary studies for example, are not so limited as these "basic" sciences and in their unreduced complexities sometimes seem to cry out for dialectical solutions.

Once in a while, in his scientific writings, Efremov did comment on his dialectical approach. Writing on his concept of taphonomy, the transition from the living state to burial in the rocks, Efremov explained dialectics briefly as follows:

> The use of the dialectical method will make application of biological data in paleozoology more fruitful. I wish to dwell on it because this method, which will in the future be substituted for 'monolineal' formal logic, is but little developed. The essence of dialectical analysis of biological and paleontological phenomena in general lies first of all in its revelation of the duality and opposition of phenomena and their development. The analysis of the development of contradictions and the unity of opposites is certain to produce good results when combined with the inevitable historical approach of paleontology.[2]

---

[2]This passage has been somewhat modified from the precise wording as it appeared in English in "Some Considerations of the Biological Bases of Paleozoology," in *Vertebrata Palasiatica II*, p. 91, 1958. To eliminate some of the "Russianisms" from it and to do this without altering the meaning, I have translated the 1961 Russian version by Efremov that appeared in Trudy IV, All Soviet Paleontological Society, Moscow, State Scientific Technical Geological Press, p. 209.

11

Frivolity, Dialectical Materialism and Science

How Professor Efremov and I explored some of the ideas touched upon in the preceding chapter will come out later in some of our conversations and letters. But while these are serious matters, they fail, for the most part, to catch the important frivolous side. Frivolity is, of course, somewhat harder to get at because culture and language are high hurdles to humor, as I had found out earlier. Only the "put down" joke seems to be universal. In the US it has been Polish jokes, Jewish jokes, and Texas A and M jokes. The Russian counterpart of the last comes out in Armenian radio jokes. Someone calls the Armenian radio station and gets an inane answer. Professor Efremov had a seemingly endless supply of these with which he regaled my wife and me when we were in Moscow together. At a lunch in his apartment . . .

Efremov:

"On a remote collective farm in Kazakhstan the wind kept blowing out candles and oil lamps at night. Children worked in fields all day, couldn't study their lessons and were growing up ignorant. What to do?

"Called Armenia radio.

"Armenia radio say, 'Simple, when is daylight, shut off room and trap light, so children can study in it after dark.'"

But he also had a more sensitive sense of the ridiculous. He delighted in some drawings of a very fine artist and paleontologist, Professor A. P. Bystrov of Leningrad. Efremov's son and daughter, like Bystrov's and most others, when young tended to get out of line. Efremov's reins were tightly drawn. To help out, Professor Bystrov sent a series of drawings he had done to keep his children in line. Some of these Efremov sent me, although they were too late to help with our brood. He told me it would be all right to reproduce them and I have done so as an example of what he found amusing (Figures 28, 29).

I sent him a "comic" book of our little swamp "people," portrayed by Kelly in "Pogo." He didn't see anything to them at all, which was much the same as my reaction to most of the humor in the Soviet magazine of humorous political commentary *Krokodile*.

Then, too, more serious conversations can take on a light twist. One time, in the late stages of a homemade feast at Dr. Chudinov's apartment, when we had gorged on delicious pelmeni—bite-sized meat wrapped in a dough blanket and prepared with a special Uralian touch—copious amounts of fruit, bread, vodka and cognac, our conversation drifted over into comparative literature.

"There are good French writers, good American writers and even good British writers," pontificated Professor Flerov in his deep voice, "but there never was a good German writer!"

General agreement around the table.

"But there was, you know," I chided him.

"Who?" came back a loud chorus.

"Karl Marx," I offered.

"He was a Britisher," Flerov came back, amid laughter.

"What do you know about Marx?" someone challenged, implying that Americans probably had never heard of him.

"I've read a lot about him and his works," I said, fudging a bit on the quantity.

"Where?" suggesting that his works were banned in the United States.

"Oh, in the books my kids brought home from school." My reply was met with polite disbelief, but no one was rude enough to quiz me about Marx. It was too much, so we dropped the whole thing.

These pleasant interludes to the daily grind over old bones were fun, and lightened my preconceived ideas of how I would get along in the Soviet Union. Very rarely, except in one-on-one conversations with Professor Efremov, did serious matters get onto the table. I shied from discussing such matters and so did

Figure 28. Sketches by Professor A. P. Bystrov of Leningrad, from a series sent to Efremov as examples of an "old time" way of keeping his youngsters in line. Bystrov was a leading vertebrate paleontologist who did outstanding work in opening up the modern field of paleontology in Russia.

Figure 29. The "maiden and the beast." More whimsey sent to Efremov from Professor Bystrov. *Dvinosaurus,* an amphibian, is from the famous excavations of Bystrov on the Dvina River in the northern Cis-Uralian part of the Soviet Union. The maiden is unidentified.

my friends. But all the while the matter of the feelings of the Russian scientists on evolution continued to plague me when ever I would put aside the comfortable feel of old fossil bones. So I put some of my questions in letters to Efremov, hoping he could clarify matters for me. After I had returned from a visit to the Soviet Union in 1961, I initiated a long interchange on this matter.

Chicago, January 31, 1962

Dear Professor Efremov:

Recently I have been reading various Russian books on evolution and species and am beginning to develop a feeling for something that I should like your opinion on. I have read, in particular, "Studies of Species" by K.M. Zavadsky and "The Theory of Sexual Selection" by L. Sh. Davitaschvili.

My reaction is that the second is heavy and ponderously written with not too much to offer and that the first is "fresher" and is written with a much more facile concept of the subject and matters in general. I have, of course, read various other general works.

Now, what seems to emerge is a very strong impression of pervasive Darwinism. Of course, all modern evolutionary thought has strong Darwinian ties, but what I seem to be feeling is something much more basic than this—the feeling of the well-spring, the inner sanctum and so forth. I do not mean that this eliminates thinking as such, but provides what all of us must have in one form or another to produce some sort of basic guiding pattern. First, my sample is small; I may be way off in my estimate. Second, if this is not entirely true, does any social pervasiveness emerge? I am inclined to the affirmative.

If you have any thoughts on this I would like to hear them. I try continually to gain a basic philosophical understanding of our science and interpretations of the implications. This strikes me as something well worth doing. In this regard, have you read "The Structure of Science" by Nagel? It has some interesting material, but as often is the case I find the biological section somewhat unsatisfactory.

Your devoted friend,
Sincerely,
Everett C. Olson

The substance of this letter did not bring a direct reply. Perhaps it was a victim of the two-way filter system that sometimes seemed to have been in operation. Most of the matters did come out in later letters. As usual, I had used somewhat cryptic and ambiguous statements to avoid censors and not to put Professor Efremov in a difficult position. He rarely failed to read in the proper meaning. What I was wondering about at this time was my general impression that the Lysenko doctrines depended on one small part of Darwinism, that which some have termed Neolamarckism, or roughly the inheritance of acquired characters. Darwin, especially in his later attempts to explain evolution, did, it is true, invoke this idea more strongly in his evolutionary thinking. In our current sense of Neodarwinism, this so called Neolamarkian concept is in fact essentially "anti-Darwin." It is this part of Darwin, some feel, that is the underpinning of the prevalence of Darwin's name in Russian evolutionary literature. As far as Lysenko doctrines are concerned this is to a large degree true. But usually it is the use of quite pure, classical Darwinism that is generally encountered. The response in a letter of Efremov's that did reach me answered my query on *The Structure of Science* in typical Efremov style.

Moscow, February 15, 1962

Dear Professor Olson:
  I was glad to hear from you in your peculiar Russian, peculiar but charming, for I really have missed you. And I must reproach you for

your enormous, magnificent and undoubtedly very expensive book you have sent me, about lost civilizations. You gave me enormous help (in your country you cannot evaluate it really) with regular sending of many and various books and now I have a much better understanding of your SF and other courses of recent literature. So don't send me expensive books and please go on with the cheap ones.

But to my horror we haven't received any copies of your book about the Permian in America and Russia! What is the matter. When have you send them to me and Dr. Chudinov and others? I have a strong apprehension that someone took your own advice and have had to steal all these books. Why you, of course, have send these books second class mail and the scoundrels steal them. So none of us have your work and I am tired by moans and groans of Dr. Chudinov, who needs your book this very moment, very badly because he now work on some fantastical compilations.

The Structure of Science I have also read but I think the author is dazzled and dizzled by the advance of physical science and mathematics. In some ways this book is harmful, especially because the author wholly lacks the dialectical method of thought. I expect many maliciousnesses from the formal logic and other formal methods because our whole world is strongly two-sided, and they haven't the skill of the use of dialectical philosophy (of course, I don't mean the so-called "dialectic" in political matters). I have become more and more convinced that our civilization with formal thought goes more and more wrong ahead to some disastrous things. But I hope before to go a long way home . . . .

About more tranquil things. Have you got A. Comfort's "Darwin and the Naked Lady?" It seems to me that you are somehow shy about such a title a search of this book is irrelevant with the Professor's dignity, but I cannot resist my curiousity.

And please, if it is possible, send me the "Photography Annual" for 1962 and 1963. Maybe it is possible to steal a copy from office or airport somewhere?

> As ever, your very friendly,
> I.A. Efremov (Old
> Efraim—grizzly bear)

The matter of stealing was a standing joke. Earlier, my wife, while in a dentist's waiting room, found a three year old copy of *National Geographic* which Efremov badly wanted. When I asked her where she got it, she explained she had stolen it, technically true. Several copies of the book titled *Late Permian of the USA and USSR* had been sent registered and finally did come back as rejected — the USA and USSR in the title was, I expect, too much. I sent them again, registered air mail, and they went through with no difficulty, only greater expense.

After this the matter of dialectical materialism simmered and died out in my work until about 1966. During this period Efremov continued his sociological comments in other contexts and some of these come out in our dialogues in later chapters. In 1966, Professor Davitaschvili, a prolific book writer, sent me a recently issued book entitled Современное Состояние Зволюционного Учемия На Западе *(The Status of Evolutionary Thought in the West)*. It was in his rather ponderous, to me somewhat Germanic, Russian, but not too difficult to read. Understanding it in places was another matter. Davitaschvili was an old-time paleontologist from Tblisi, Georgia, and immensely widely read. We met just once for a short while. After a quick go-through of the book, I dismissed it as more or less trash. On the second reading, I began to wonder what prompted all this "nonsense." Within days of delivery of the book from Davitaschvili I received a second copy, from Efremov. This prompted me to initiate a long series of letters on the matter, for I had decided to comply with Davitaschvili's request that I review the book.

Chicago, September 28, 1966

Dear Professor Efremov:

Not long ago I received the two books and pamphlet that you were kind enough to send me. Before I received the books from you, Professor Davitaschvili had sent me his book on "The Current Status of Evolutionary Thought in the West." He asked me to review it. So I have been reading it and have just about finished. Naturally this will not be a book that would be too popular here or in much of western Europe. That is fine, for we need much more interchange of different concepts. There is no question that there is a great deal of local ingrowth in all areas in all fields. The immediate reaction to the book here, I am sure, would be "Oh God, that sort of nonsense!" It would be put in the same lower drawer as Goldschmidt's "hopeful monster" ideas and so forth. This is not because Darwin is not very highly regarded here, but because of the general concept that there is one and only one way that evolution can be understood and that this way has implications that go far beyond objectivity, which, oddly, is the same criticism that is made much of in what is thought of as the West by Davitaschvili.

I don't react this way since I have a "failing" of open-mindedness, which irks some of my associates. I tend to see both or many sides of questions. Now I do want to review this book and will do it. Just in what manner I don't know just yet. Probably it will be a factual account of what is said. Yet it will certainly be necessary to include some commentary. It is in this area that I am somewhat unsure.

I don't know if you are in physical shape to answer some questions for me. If you are not, please don't even think of trying. I am writing on the chance that you are for you are the only one I know who can give me objective opinions and who is fully aware of the issues involved.

Question 1. How well thought of in Russia is Professor Davitaschvili? I have met him and found him pleasant and have read a good deal of his work. He appears to me to be a widely read, paleontologically oriented biologist, one who probably has certain deficiencies in contemporary evolutionary studies, as they pertain in particular to recent DNA and RNA studies. I respect him highly but he seems to have something of the aura of the grand "oldtimer." How far off am I in this judgement? How do his contemporaries and the younger men in the Soviet Union judge him?

Question 2. My contemporaries in this country and some parts of Europe felt the impact of Lysenko very strongly. I note that there is almost nothing in the book relating to this, which seems to have had a very strong influence in your country. Is this because these concepts are now largely discounted, or are there other reasons?

Question 3. This is the most important and most difficult to phrase. In later parts of the book, the basis for criticism of the "synthetic theory", or his "Post-Neodarwinism", becomes clear. In being based on mechanistic materialism, Davitaschvili states, this theory (the Synthetic theory) cannot encompass in simple terms some aspects of evolution. In particular it finds difficulties in interpretations and explanation of "directionalism" and in the face of these difficulties, finds it difficult to avoid a finalistic or idealistic approach. This is my paraphrasing of what I think is said in many places in the book, but made clear in next to the last chapter. Darwinism, which is somehow equated with dialectical materialism in fundamental philosophy[3], can be understood only under this philosophy, as I interpret what is said. Classical Darwinism has all the answers to evolution.

Now I am aware that Darwin was knowledgeable with respect to Marx, but I find it hard to see that in his class position in the social structure of England during the early and middle portion of the 19th century the germs of dialectical materialism were in his own work. Now then, is it to be understood that his work can be fully appreciated only under this philosophy, which began in his time, but certainly

---

[3]In a "post-Lysenko" scholarly book, N. P. Dubinin, whose work was suppressed during the time of domination of the Lysenkoites, makes this clear, contrasting Darwinism as strictly materialistic dialectic with Neolamarckism, which is considered nondialectic (Dubinin, N.P., *Evolution, Populations and Radiation,* Atomizdat, Moscow, 1966, pp. 190–191). The book came out and Efremov sent it to me as this correspondence was going on, but I had not seen it at the time of this letter.

came to flower and fruition much later than when Darwin was maturing his ideas? I am asking this as a serious question and one on which I would like your opinion.

In this same vein, Davitaschvili says that western scientists, while contributing enormously to factual aspects of biology, cannot entirely appreciate evolution because they do not understand dialectical materialism. Now, any criticism that I might raise could be countered in this way: that I don't understand the philosophy and thus can't be a critic. Probably I don't, although I have read as widely as time has permitted in attempting to do so. Could you very plainly indicate to me in what manner this philosophy very objectively gives an explanation of directionalism that does not in the end fall into the same problem of ultimate explanation as any other explanation, such as that of mechanistic materialism? This is probably a big order. But it is critical and I don't feel it is answered in the book.

Question 4. This gets to the final question, call it 4, although it relates to the last. I quite agree as will most scientists in this country, and as Davitaschvili recognizes and emphasizes, that no aspect of metaphysics can logically be introduced into scientific work per se. Where this is done in explanation of evolution, we have departed from the realm of science. I quite agree, as well, that any dogma which limits the realm of operation in study is inhibiting. But, in a somewhat dialectical way, I have just reached the heart of the problem. If there is any one way to truth, this way is necessarily dogma, whether it depends on some metaphysical guidance, some intuition, or some series of basic premises empirically derived, or some set of independent a prioris. In one way or another we must arrive at an explanation of our framework, be it one of order or one of chaos. If we arrive at it and consider it the way, then how does any single basis escape being dogmatic or becoming dogma?

As I understand it dialectical materialism finds its base in a particular set of relationships that are in essence basic premises which, just as the order of which I have spoken, may have empirical justifications but which similarly cannot be explained by the simple statement "that is how it is." In other words, at some level there must be a set of undefendable premises or axioms. All that we do thereafter, empirical observation included, depends on these axioms. As far as I can see all systems of thought must have some such base, a base beyond which they cannot penetrate without violating the limits set by the premises.

Now it seems to me in suggesting the various ways of thinking about and studying evolution that Davitaschvili has taken up, all must fall within this context. Those that desert a materialistic base go beyond in a metaphysical abandonment of science, which casts them beyond the pale of scientific investigation. When a vitalistic or finalistic tendency creeps in, this is what it does. It arrives at the time when the limits of explanation, under the premises, have been exceeded. Specifically with regard to directionalism, since it has been a puzzle under synthetic theory, some have certainly tended in this direction.

But how does directionalism under a dialectical materialistic philosophy avoid this tendency, if it does not go beyond the basic premises in which, as I understand it, a materialistic basis for changes in direction, specified or unspecified, exists? I am perhaps wrong, as Davitaschvili would suggest, in my understanding, but does not the social implication of this philosophy involve not only a directionalism but a very specific and infallible aspect of this directionalism?

As you can see, this is asked in good faith. I am most anxious for an interpretation from one who knows this area well, as I know you do. It is not in this sense an impertinence, which it could well be considered if you did not know me well.

With deep regards,
Your friend,
Everett C. Olson

A note about the "Lysenko Affair" may be in order here. Lysenko was an "agriculturist" who developed some "new" concepts to hasten the growth of crops during years in which the food crisis was severe. These were basically crude and depended, as they were developed, both upon spurious "genetic" theory and to some extent falsification of results. This work, with the eventual help of a "state philosopher," was given an aura of correctness under the superimposition of dialectical materialistic concepts. Through a series of political manipulations in the 1930s and 1940s the concepts gained the full support of the Presidium and of Stalin, and became the prevailing doctrine, not to be countered. The upshot was, in effect, that not only did agriculture suffer seriously but the field of genetics, which had been strong in the Soviet Union, went into eclipse. The best of the biological scientists were, in very large part, excluded from science and many were "lost," some left and some clearly died in concentration camps. The name of Michurin, an obscure horticulturist, later became applied to the doctrines—as Michurinism. Only in the mid-1960s did an effective recovery begin.

But back to my letter. To ask for answers to all of my questions was a big order. I am sure, had I written inquiries along similar lines some years later, I might have taken a somewhat different approach. The letter, however, elicited a very long reply which I am including in its entirety because it brings out so many things critical to Efremov's whole consideration of the role of dialectical materialism, things deeply ingrained in his thinking.

Moscow, October 5 [sic], 1966

Dear Professor Olson:

Your second letter, date October 7th [sic], arrived with amazing speed (however quite normal for normal circumstances) after a week delivery which I believe is an indicator of warming relationships. But first letter, dated September 28, arrived only yesterday.

You must not feel guilty with your questions. I shall answer you much more questions if I will be able to do so.

Question 1. Prof. Davitaschvili is far off the whole appreciation here. During the thirties and forties he occupied a very orthodox position in the line of the regime and Lysenko and bring many difficulties to the more broad-minded paleontologists. Thereafter we named him "Davite shval" [smash the trash is the exact translation], for clear flunkyism. But, of course, he is a very "vast reading man" and much more acquainted with world literature than any of our other paleontologists, especially on generalized books and generalized pamphlets. From the other side he is only a cabinet worker (office worker) and knows very little about field explorations and geology as real geology. You have truthfully mentioned that he also is outmoded in all new horizons open by new branches of biology with applications of molecular biology, biochemistry, geochemy and many others.

Partly in answer to your second question, L.Sh.D. was in the late forties an eager follower of Lysenko's dogma but now has written several pamphlets against the latter in the name of Michurin, who as a creator of the new path in biology is quite a mythical figure I believe and when lived never pretended to be more than a good selectionist as L. Burbank, for example. By the way how many perfect old sorts have been awfully damaged by unskilled introduction of the new ones and enormous doses of ignorance among Michurinists?

Question 3. You perfectly clearly have formulated the essential in the position of the line of criticism in the book in question. The dialectical point of view of Davitaschvili's book on my sight is very weak and formal, being formulated only in words and also seems like a mystical force.

The "directional" way in your use of terms (orthogenetic is old name I am more accustomed) is in my opinion necessary for every deeply experienced paleontologist because all the materials in our hands cannot be explained in another sense. But among all of our scientists (Davitaschvili included) it seems to me, boldly said, that only I endeavor to explain the orthogenetic way of evolution in dialectical sense. The others openly ignored such enterprise and mentioned the "dialectical materialism" only as a general "Word." It seems to me that the general physical environment of life act as a "corridor" whose parameters are the "limitized" and "pushed on" forces of the evolutionary process as a whole. The goal, because the essential feature of living organism is the constancy of inner conditions (homeostasis) without which all heredity and the work of the biological machine is impossible, is the freedom from the environment as widely as possible. More freedom — more storage of information, etc. This struggle for freedom if regarded integrally is aristogenesis by Osborn or aromorphosis by Severtzov. The orthogenetic (directing) corridor of general physical environments is nomogenesis by Berg if regarded as the only possible way in evolution (integrally). To the Neodarwinist the adaptability of evolution to the environment as well as the selective process leading to complexity and high fitness is clear, but without the "bridge" to the general target and therefore without understanding of the mechanism as a whole process. Really I believe this "mechanism" is quite a dialectical one: the necessity <u>accomplished through the sum of causalities</u>. Necessity here is the freedom from environment and causalities are adaptations.

Therefore in organic evolution we have a plan (a predestined one) instead of blindness (or rather randomness) of the process itself. <u>This is also a dialectical point of view</u>—<u>the two sides of the whole or the unity of the contrary</u>. Of course, the historical chain to the general goal ended on earth in Rezent, but such is our scientific way of collecting knowledge. I doubt that this is quite understandable to you in my inadequate English, but if you understand the general trend of thought it may be of some use to you.

In the Davitaschvili book are only recommended for "western scientists" to understand dialectical materialism. But to accomplish such understanding through quotations from Marx or Engels is a hopeless task because the very "matter" today is infinitely more complex than a century ago. I believe you may openly criticize this recommendation, however, the clearance of "<u>weak</u>" points in western theories is very useful for every paleontologist and for this target Davitaschvili's book may be highly recommended. But as for understanding the method of "dialectical materialism" the author accomplished none on my view.

I believe that recently only three evolutionary students whom I appreciate very much, you, A. S. Romer and G. Simpson. The latter is, strange to say, very similar to Davitaschvili only on the other side of the line of truth. He, as many formal thought scholars, ignores all opposite events if they appear beyond the field of narrow investigation. Davitaschvili ignores all opposite facts if they cannot be packed in a prematurely generalized trend. From another point of view "dialectical materialism" is a very old philosophical trend, which in occult books regarded as the "Great Mystery of Duplicity (or Double)." The attempt to regard every fact or event as a sum of two opposites from the two sides in the same moment is of course much higher than formal, linear logic but infinitely more difficult and therefore possible only for outstanding minds. The dialectical line of thought also includes historical aspects of all events and this is the cause that the paleontological and geological investigations appear more dialectical than other sciences without the historical basis.

The cardinal difference between dialectical materialism and dialectical non-materialism lies in the "Primo Motore" [i.e., the general cause of events]. If the "Primo Motore" is a result of pure material events in accordance with general physical laws, then this view is materialistic. If "Primo Motore" appears to be unrecognizable and somehow beyond the general laws of the material world then it is "vitalistic," "reactionary," and "unscientific."

One more example. The very wide use of Marx's formula: "existence (mode of) determines the conscience" in this form is really a metaphysical one, because it lacks the other side: "conscience determines the mode of existence." Now it comes to the Marxists very slowly that spiritual conscience is a quite real force especially in the fitting, survival and "way up" in wholly materialistic processes. By the way, if spirit is the highest form of matter, what then? Why cannot it be a real force and inevitable "other side" in a dialectical world?

As you can see, all this agreed with your considerations about the question 4, because true scientists cannot operate with any dogma, because axiomic answers to all questions are religion, non-science. A scientist also needs "homeostasis" but only adaptable to the rapid alternated knowledge and if knowledge came upon the exponents this adaptability must also be rapid ones. Therefore only direct explanation of the discoveries by the dialectical way of thinking has a scientific value. So one must have very good brains to be able to do so. If not, we must bear with the formal one-sided views.

Between us girls, I have two heretical views for the professional scientist. The whole process of obtaining knowledge is dialectical (two-sided). Scientists on the one side explain the new by the orthodoxal old. Others (on the other side) explain the old and new facts and events by the invented and mysterious "metaphysique." Both compose a unity of polar opposites and our knowledge progressed between them as I attempted to show in my romance   [*The Razor's Edge*]. Only the strong minds have ability to find this razor's edge immediately.

My second thesis is this: the universe now appears so infinitely complex that we can discover <u>everything</u> [note, I believe he intended *anything*], and can predict many discoveries! We can predict correctly really all if we only formulate the event with a satisfactory clearness and within the general parameters of the physical universe. It is similar as to cutting something from a snow wall—we can master every figure desirable from the cube to Aphrodite. So I haven't valued the so-called "predictions" articles in recent science because there are millions of this stuff and success with one or another prediction not at all verifies that this way of investigation is the only truthful one.

As a result of unexpected (expected by dialectical philosophy) complexity of the universe is that the formulation obtained became more and more undigestible and useless. We drown in the deep ocean of facts and experiments and the proud tower of science more and more became like the Babel tower. Gradually ruined from within by utter ignorance of the scientists themselves, 240,000 pamphlets in chemical science every year, 90,000 in physical and so on and speedily up! We are the last scientists in the good old sense of this word . . . . But enough of this babel! You must have a strong head, I fear to be able to read such stuff.

Resume. I think your opinion about this book is a correct one. It must not be easily laughed off because of contain useful critical aspects on many "theories." But the recommendations given are weak because of too generalized a formulation which is of no use on concrete material.

I have just sent you three books. One by Semenov, about the origin of Man is essentially typical for Davitaschvili's line with much more flunkyism and citation. You may find it interesting as example of the newest and serious book which is really an "old song" dogmative and metaphysical, entirely abandoning the second level of evolution of man — spiritual. The "herd" — this only he sees on the early stages of evolution.

Your friends as always,
I.A. Efremov

P.S.: Please let me know about receiving this letter as soon as possible!

Chicago
December 2, 1966

Dear Professor Efremov:

The Komsomolskya article on the "Dialectic in Science" arrived and is most opportune. I am still stewing about the whole problem, the problem of how I and my colleagues of the "west" really think about evolution. We are not consciously dialectical, but in a way I don't know how this subject can be otherwise conceived, for surely there is a duality and it would seem that what is critical is whether or not this dualism plays a role in thinking or whether there is a dual linearity with little cross-relationship. I think that the whole matter is in no way simple and that categorization of an area of cultural continuity as this way or that loses sight of the variety present and that, furthermore, this variation goes right down to the individual as he thinks on one thing or another.

Thanks for the new books you sent me. The one on the origin of man sounds a lot like some of our anthropologists but with a different dogma in the background. It seems to me that it is characteristic of many who work in this area to take some simple, conceptual framework, believe it works without question and base everything on it. It builds some mighty towers but I have an uneasy feeling about the base. Of course, in a way, all science does the same, accepting some basic axioms and some limitations, but beyond this there is an empiricism which I miss in most writings on man and society.

When I do finally finish up what I am trying to write on the matter of Professor Davitaschvili's book and others related, I would like to send the paper for comment. Do you think this would be a good idea and do you have the time and energy to look it over? I would not need the copy back, but would like your comments. If you think it is a good idea, please say so.

With my best wishes,
As always, your friend,
Everett C. Olson

Moscow, January 8, 1967

Dear Professor Olson:

Now to answer your questions:

1. Of course I will be glad to look over your article about Davitaschvili. At any rate send me only a third or fourth copy to void the evil of no returning.

2. I agree with you completely on your evaluation of Semenov's book. The same, half-fanatical tendency but on diametrically opposed grounds.

3. My "Razor Blade" [*The Razor's Edge*] you cannot obtain. I like your idea of a dialectical trend of thought as two intercrossed lines of opposed linear thoughts. By the way, my taphonomy is based on the opposite side of sedimentation of the geological chronicle—the process of destruction. In other words—the other side of the medal, as we say here. It is very interesting to take a look on the other side many of your forebearers may have overlooked.

As ever, cordially your friend,
I. Efremov (Old Efraim)

This letter was the last of the series related to *The Status of Evolutionary Thought in the West* by Davitaschvili. In his letters Efremov, of course, was dashing off his ideas as they came to mind, without the caution he would have used in publications. In his fictional writings, some of his unorthodox ideas are better expressed through the vehicle of his characters in other times and other places.

I did finish the paper and sent it to Efremov. He made some corrections and predicted "some furs will fly." The paper appeared in *Evolution* as a long book review. In it I tried to explain dialectical materialism and did not, I now feel, emphasize sufficiently a main theme of Davitaschvili, that Synthetic evolutionary theory, or what he called Post-Neodarwinism, could end up in idealistic thinking. Idealism, of course, is a complete anathema to Marxist doctrines, an abstraction that alienates explanations from the basic materialistic reality. The Marxian "flip-flop" of Hegel bore precisely on this point. Efremov brought this out in noting the "lack of a bridge" between the major aspects of Neodarwinian theory, the "random" changes through mutations and the directing effect of natural selection. This now seems to me to have been the main theme of the book and what Davitaschvili was writing about when, in a review of my review, he complimented my effort to understand dialectical materialism, but noted that I had been only partially successful.

Having obtained some understanding of this point of view and the complex web of Efremov's thought, I found my perspectives enlarged and my skepticism about the completeness of the explanatory aspects of Synthetic theory somewhat increased.

My rather mild worries about the sufficiency of the Synthetic theory did not arise in a dialectical context. Rather, some aspects of the fossil record did not seem readily explainable by it, a position held by others but rarely expressed after the middle 1940s except by a few "radicals." I expressed my reservations in a paper presented at the Darwinian Centennial at the University of Chicago in 1959 ("Morphology, Paleontology and Evolution," in Sol Tax, ed., *Evolution after Darwin*, volume 1, pp. 523–545). I had been asked to read all the papers, as someone who was at hand as they arrived, and write what I thought was needed. I did in a sort of "whoa, let's step back and take a look" paper. The synthesis was in its heyday of popularity at the time.

My paper was not enthusiastically received. The problems of adaptation lay at the base of my skepticism, involving such things as: (1) the existence in the past of seemingly "adaptive monstrosities;" (2) vast parallelisms in development of morphological complexes in evolutionary lines only remotely related; (3) mass extinctions at various times in the fossil record, involving many different kinds of organisms; (4) long temporal sequences in fossil lineages in which major changes are persistently directional, often seemingly nonadaptive in their initiations and with no evident environmental stimulation. As George Simpson said to me, rather pleadingly, all of them *can* be explained under the synthesis. It does require reaching pretty deep. Since 1959, some points have been resolved, but most are still puzzling to me.

Simplistically, the synthesis involves natural selection of variants in a population. The variants result from differences in the genetic constitution of the immediate forebears, by mutation or recombination. Selection preserves the most fit (survival of the fittest) and thus modifies the genetic composition of the succeeding generations. Slow accumulation of changes set the pace of gradual evolution. There can hardly be doubt that this is one aspect of evolution, but the question of whether or not it is a sufficient explanation of all change remains and has been of increasing concern to some evolutionists during the last 15 or 20 years.

Dialectics, as Efremov applied it, could allow a fresh look at some of these problems, a look he sketchily outlined in his letters but never formally published. Where, then, does this leave matters? I spent many hours during 1971 talking with Professor Efremov about these and other matters. I had hoped to continue in later years, but his death in 1972 precluded this. As far as science is concerned, only added emphasis and some clarification of the ideas expressed in his letters emerged. He noted that in his opinion about 99 percent of the Soviet scientists did not follow dialectical materialism, although it was such a significant part of their education that some of its tenets probably remained in their subconscious. About one percent of the scientists, however, did accept dialectics, and these, with a few exceptions, did not really apply it properly. The tragic and dangerous situation of the world, which Efremov felt strongly, was due in some large part to the dominant position which linear logic had achieved.

Dialectics, he maintained, is not a law, but an expression of the basic structure of the universe, and thus all evidence, necessarily materialistic, must be viewed as dialectic. Both formal religion, in which the "reality" is an abstraction of God or the equivalent, and the materialistic so-called "reality," are harmful to science and hence to mankind. Once the truth of the concept of contrasting and interacting opposites is understood, the great problems of science—that is, the basic conceptual problems—begin to fall into place. Man has emerged as an intelligent organism by a dialectical spiral of evolution in which the balance of physical and psychic, both materialistic, have interacted. Wherever intelligence has or will emerge in our universe, it will be in the guise of creatures morphologically similar to man, but suited to their own particular environments. These were his main thoughts, greatly condensed.

Many persons, of course, do not go along with this and Efremov fully recognized one's choice to see things otherwise. Dialectics represents, as he has said, a way of looking at the sciences that have historical contexts, geology, paleontology, astronomy and so forth. I had found it strange that so few scientists in Russia operated within a dialectical framework in a nation where this was "state doctrine." I was surprised also that those few who did seemed to be in large part sterile. One simple explanation is that "hard" science is necessary, and because symbolic logic and consequent mathematics are necessary, even though "linear," the basic dialectic philosophy must be put aside to accommodate these linear essentials.

An interesting final footnote to this matter of dialectics is to be found in the volume that resulted from the First Biometric Conference in Biology and Medicine, sponsored and directed by Professor P. Terentjev (1961), a Leningrad herpetologist. This includes an essay by the State Philosopher A.E. Popov. Biometrics, dealing with probabilities and logical mathematical solutions, was slow to develop in the Soviet Union, even though some of the world's greatest mathematicians have been Russians. The "philosophical" statement legitimized the work of the participants in the conference, or rationalized it, if you will, as follows. Full reality exists only in the temporal context of dialectical materialism. At a given time, however, with the flow stopped, so to speak, a powerful solution of immediate problems is to be found in a mathematical, statistical treatment. This has a strangely Bergsonian flavor. The "slice of time" reveals only partial reality. What emerges is not total truth, but heuristically is an appropriate treatment of limited problems, therefore perfectly legitimate.

12

The Other Side of the Medal

A Dilemma

Efremov's ideas had piqued my curiosity. At first his thoughts and philosophies did not conform at all to what I had expected in view of my perspectives on the society in which he lived. But thoughts are doubly filtered by the mechanics of communication and then by preconditioned interpretations.

Some bits of Efremov's philosophy have come through in his scientific writings, although mostly somewhat masked. They rise nearer the surface in his fiction, but here there is a touch of missionary zeal. His letters and our long dialogues were more revealing, even though they were based on thoughts of the moment and contradictory from time to time.

What came out of all of this, especially our letters and the talks, was not at all a bold set of unified principles. Rather there was a sense of confusion and vacillation, which seems symptomatic of so much of modern intellectualism everywhere. "Where are we going?" was the underlying theme, the basic dilemma—somewhat masked by his brusque Russianism, but never buried. Efremov reached far beyond the current troubles of his own country to those of the world and sometimes of other planets.

Most of us are, I believe, more narrowly concerned with today and tomorrow. When we in the United States in the 1960s, and until very recently, talked casually about the Soviet Union, we rambled on about repression, anti-Semitism, Godlessness, a Russian lust for war and conquest, the worldwide spread of communism and like matters. The pseudo-paranoia of the early 1950s gradually abated, but continued to lie in a shallow, eroding grave even in the 1980s. How do the Russians think and feel about the United States, its people, and its government? From my limited contacts, I would guess that there are as many ideas as there are groups of thinking people, and less consensus than here. But I have no business saying how the Russians feel about us and the world and about their own roles in life. I leave this to the experts whose business it is to know. Yet, it was in this general web of complex thought that Efremov had grown up and lived and, starting from it, I was able to gain a sense of the thinking of a small fragment of the intellectual whole, and particularly of Efremov.

As a science fiction writer, like most of his kind, Efremov was concerned with the future. Most of us think about and write about it more in the short term than the long, but he wrote about it, talked about it and worried about it as a long range continuum and repeatedly made contradictory judgments of what was to come. He was seriously concerned about, and voiced a moral responsibility to, future generations. He sensed doom. His deepest conviction, probably—"We better both be afraid of the Chinese." The cycles of Indian and Oriental philosophies plotted still another course, with danger at the peaks of swings and no assurances of safe passage. But the balance—on the razor's edge, the dialectical power of opposites—promised a way out. "Which will it be?" was the unanswered question we revisited many times.

Ways to Go

Downhill

To the science fiction writer, even though in the future the earth must go, perhaps in some sort of holocaust, life must go on elsewhere and, for intelligible story content, it must be in the hands of humans, or humanoids or their creations. Other planets and stars are havens; physical laws, often badly bent, are the constraints. Usually the baser side of humanity—avarice, wars, conquest, deceit and cruelty—persist.

Not so for Efremov. Although vigorous and physical, as a geologist in the field or a sailor on the sea, he was a most gentle man who abhorred violence. Gymnastics and ballet were his favorites, and contact sports offended him. He found most distasteful the physical singing of "Sachmo" (Louis Armstrong), his popping eyes and sweating. Humans, he was convinced, in their arts and sciences of the brain and spirit, have the capacities to master violence and aggressiveness by formulating societies in which motivations for such behaviors had no reason to exist. This is a constant theme of Efremov's science fiction, and it had wide appeal to his fellow Russians. I can't help but feel that this was mostly cast in a mold reminiscent of Dorothy's dream of the Land of Oz somewhere "Over the Rainbow." To Efremov, I found over and over, it was real, and it was his mission to tell it to the Russian people in whatever way he could.

Yet, in 1969, when my wife Lila and I were spending a pleasant afternoon in his apartment on Gubkine Street near the University of Moscow, Thais, or Taya, Efremov's charming second wife, brought us a diagram. Efremov had constructed it based on prophecies of ancient Indian and Tibetan seers, and Thais had drawn it up especially for us (Figure 30). He explained it as follows.

> The things are somewhat gloomy in this world, for the near future especially. This is a coincidence with the old Indian and Tibetan prophecies about minor and major peaks. I have drawn them graphically on a diagram. A down peak in 1972 [this was 1969], a real up peak in 1977, and a very big downfall with gigantic wars between 1998–2005—age of the White Horseman of Maytreya. But I haven't a chance to reach this age, maybe you?

Unless this dangerous time could be passed somehow, our ancient civilization was ended. Earlier, in 1966, I had written with some misgivings about where we were headed.

> We here (in Chicago) are in the midst of a very hot and humid summer, and, as they say, "the natives are restless." There is tension in the air. It's all part of growing up, but there is such a long way to go. The world is always somewhere in a state of revolution and this century is one of accelerating pace with so many adjustments that it is little wonder that Homo sapiens is having a hard time keeping up with the pace. It is as if the whole world is trying to drive some complex modern freeway road system that requires repetitive split-second decisions, moment after moment, in order not to crash, or at least to avoid going round and round on a butterfly.

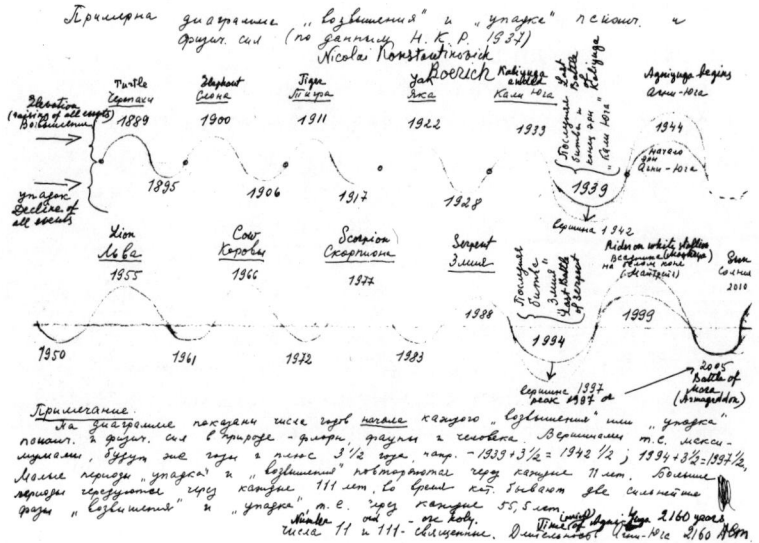

Figure 30. The cycle of the years. Times of rise and decline of humanity to Armageddon, the Battle of Mora, at far right. The period of Agui Yuga lasted for 2160 years. From 1994 and the last Battle of the Serpent, time continues to Armageddon in 2005. Drawn by Thais Efremova from sketches by Efremov. This figure has been reproduced from difficult, informal copy, a gift from the Efremovs.

Efremov replied,

But I agree with you completely that in the second half of this century our species is not only having a hard time keeping up, as you say, but trying desperately to find his place in the new and not very palatable world which emerged around. I, at my own, having a sharp memory, can see clearly the way things are at the beginning of my career as a scientist and now. An awful difference. To begin, the scientist is no more a free searcher for knowledge, but only a highly qualified government worker as well conditioned as the others. We paleontologists enjoy some last freedom for the price of neglect and absence of "honour." But it shall not be for long. With the environmental dangers to our genes' pool and fast extinction of many plants and animals the interest in paleontology must revive by the end of the century, and people will make memory to all of us (if they only understood).

The thought that the values of history, and especially of life history from a naturalistic point of view, were being lost was a constant and severe irritant to Efremov. One more comment along these lines came in response to a simple complaint that I had made about a boy we had hired to keep our garden alive while we were in Russia in 1971—and who didn't.

Efremov, in a letter on many matters, went on:

> One more point. Disaster with your garden because of a "lousy" boy is a very abundant event now, and I guess throughout the world. The unreliability, laziness and naughtiness of "boys" and "girls" at every sort of jobs is characteristic of this very time.
>
> I call this the 'immorality explosion' and it seems to me more dangerous than nuclear war. We can see through very old times that morality and honour (in the Russian sense of chest) are much more significant than swords, arrows and elephants, tanks and dive bombers. <u>All</u> destruction of empires, states and policies came through moral depravity. This is the only real cause of ruin in all history and therefore destroying is the only self-destructions we became aware of with nearly all diseases.
>
> When all men accustomed to honest and hard work [have] passed away what sort of a future awaits mankind? Who can feed, clothe, cure and transport people? Dishonest, as they basically are now, how can they do scientific or medical investigation?
>
> Generations conditioned to the honest way of life must be extinct during the next twenty years and then a greatest disaster in history will come of widespread technical <u>monoculture</u>, which basically now persists in all countries, even China, Indonesia and Africa.
>
> Have you ever heard of the book by Alan Seymour 'The coming self-destruction of the U.S.A.?' It is issued in England in Pan's paperback, but I cannot afford it and don't know the author's meaning. Maybe you can see through it. Is it good or worthless warning?
>
> But I must end this prognostication and wish you and Lila with all our heart from both of us good luck and health.
>
> Your always and fond friend,
> I. Efremov (Old Efraim)

This was written in 1969 and, to date, has been quite prophetic. Probably some of both his and mine is just 1960 to 1970 "generation gap" carping. At least to some extent, but by no means altogether. That his sense of youth and morality in the controlled Soviet society was cast so strongly, and that he saw the consequences as so disastrous, startled me. When I entered into all of this in the late 1950s I had thought that the regimentation in the Pioneers, common mandatory education and common language throughout the Soviet Union, collectives in the country and organized apartment complexes with local "commissar" supervision and meeting quarters, would have resulted differently. Not in a way I would have liked, but surely not as Efremov viewed it. An avowed communist in a broad sense, he saw the beginnings of Soviet immorality to have been in the Stalin purges which "killed off all of the intellectuals" and left a serious vacuum amidst the uneducated and unmotivated. Where, he so often asked, can the intelligent and devoted teachers come from?

Cycles

The spiral of dialectical progress is reputed to be the way of history. Efremov talked of it and expressed a strong belief in it, but his attachment to linear cycles, seen in his love of prophecies, was at least emotionally strong. I didn't fully realize this at the time I sent him a copy of *A Canticle for Leibowitz* by Walter M. Miller, Jr., first published in 1959 and now in at least its 20th printing, but it touched base. Oddly, what was to me a basic theme of this brilliant science fiction novel, that of cycles, never seemed to impress him as much as I might have expected on the basis of what I later learned.

A Canticle . . . was cast in the 32nd century long after the earth had been devastated by a nuclear holocaust. The setting was a monastery in the desert of western North America where a young novice led the way to preserved fragments of documents from the 20th century with the name of Leibowitz distinguishable on them. An ecclesiastic search led to interpretation in context that preserved ancient formalistic catholicism, and Leibowitz, an obscure technician of the past age, at length became the saint of the monastery. The main theme was the repeated realization of the tendency of man nearly to exterminate himself and then, reeling, to build once again on the scattered ashes toward a new and more proficient round of destruction by an ever more efficient technology.

Running through the desert scenes of desolation, despair and genetic upheaval, the rise of new technology and the persistence of ancient, formalized mythology, is the trail of an enigmatic, knowledgeable stranger. It was he who pointed out to the novice the way to the Leibowitz documents during his Lenten fast in the desert, and thereby stirred the life of the dark world. At that time the stranger was a gaunt traveler, a pilgrim, loins girded with dirty sackcloth. Later he was a hermit, the keeper of a strange goat. Somehow, the figure of "Saint Leibowitz" in the monastery reflects his gaunt, slyly grinning face. Searching, still not finding to the end, he emerges as Lazarus.

Finally, technology rises supreme again, as mobiles rush by the monastery and flying machines dominate the air. Armageddon, once again, is witnessed by the chosen few from the monastery as they leave with the ancient treasures, bound for the stars in a long-readied spaceship.

> The visage of Lucifer mushroomed into the hideousness above the cloudbank rising slowly like some titan climbing to its feet after ages of imprisonment in the earth.
> The wind came across the ocean, sweeping with it a pall of fine white ash. The ash fell into the sea and into the breakers. The breakers washed dead shrimp ashore with the driftwood. Then they washed up the whiting. The shark swam out of his deepest waters and brooded in the cold clear currents. He was very hungry that season. (From *A Canticle for Leibowitz* by Walter M. Miller, Jr., J.B. Lippincott Company, 1959).

Late in 1960, Efremov wrote me his initial reactions:

> Thanks a lot for the books. Now all arrived safely, A Canticle and a heap of S.F. magazine. I have read the "Canticle" and enjoyed it very much. Very interesting and wise book and I see your point [I am not sure what my point was, probably about the repeated resurrection of man]. But for my taste it is somewhat without the "wine and colors of life" and narrow in views on nature—undoubtedly in the jewish pattern.

The whole Jewish matter—in Russia where he knew it firsthand, and elsewhere—was much on Efremov's mind and surfaced often. His was a mixture of admiration for ability, respect for "good Jews" and fear and dislike of "bad Jews," the money changers. In his somewhat obscure days between 1915 and 1920 he had been in the Ukraine and had grown up among Jewish peoples. This he never mentioned to me except very indirectly in his remarks on Leibowitz. I asked him from time to time about what to me was the irrational suppression of Jews in the USSR. He rationalized it as follows. The Jewish youth, being southern, mature more rapidly than the northern Russians. Many are extremely intelligent, but ruthless. If all top positions in government and science are not to be taken over by precocious Jews in Russia, which he deplored, it is and has been necessary to set up quotas, to deny freely competitive admissions to higher education, to positions in institutions, and to maintain a balance in political offices. Progressive malcontents and troublemakers inevitably resulted from this unfortunate set of circumstances. Were free emigration policies in force, there would be an exodus and a severe brain drain, which the country could not afford, particularly after Stalin. Only in an ideal communistic state would such problems vanish! I have paraphrased his thoughts, but these points came up many times, directly or as undertones such as his remarks on the "Jewish pattern" which must be searched out at best, if it exists at all, in *A Canticle* . . . , except by inference from the characters of Leibowitz and Lazarus.

Later, Efremov wrote more on *A Canticle* . . . :

> Some new considerations about "Canticle" I think the general feelings of you and the author about the future are correct, but not on Christianity. Now this is an idea! Our future may be analogized with the early centuries of the Christian Era, with the correction that the general advance is greatly accelerated, when beautiful, wise and wide philosophy, art and opinions of the Antiquity have been completely revised by the new, more gloomy outlook. After this collapse the dark ages came inevitably. But what was the cause? Only one—the promise of equal and good life for weak people! Now if such weaklings' percentages are great, then inevitably these opinions must seize the whole world and the only solution is to confront them with similar doctrine, as Buddhism of that time. Now after the "Gay nineties" we all stay once again in front of quite similar circumstances. The new doctrine of promises of good life for every man in enormous masses of population shall inevitably seize the whole world and the necessity haven't anything to do with such a force as in times many centuries past! It is only hope, I think and between us girls, I put myself a question: shall I see the Dark Ages II or die before? Lupum auribus tenere—an exceedingly good Latin proverb that means one doesn't know what to do in front of danger.
>
> The 2nd Dark Ages, in the "Canticle" of course, was but one in a long, never ending succession for our species.

The content of this letter was a little vague and puzzling. In still another letter, also obscure and with pertinent passages too scattered to quote verbatim, Efremov strongly made a point, based on his experiences in the first half century of Soviet Russia, that the formalistic aspects of religion could not survive a holocaust as suggested in *A Canticle* . . . although religious spirit might. Formal religion was, he averred, virtually destroyed by the Russian Civil War and during the Stalin regime. Thus, that a monastery and the fabric of Catholicism could have persisted through and after a holocaust, he denied. I replied as follows:

> On Leibowitz, I quite agree that this is unrealistic from your point of view. It does seem unlikely that formalism and structure could survive the holocaust little changed. This didn't upset me in the story, for it seemed a simple vehicle for the underlying concept of the massive "will" of man to surge back after self-inflicted tragedy. It was contact with long-past reality (our today) by direct comparison and continuity rather than by change and contrast. I do think the author likely was strongly influenced by the preservation of culture in the later parts of the Dark Ages through monasteries. Of course, his analogy does not admit a place for the preservation of other cultures existing during the time of rise of the institutions that later took over. If nothing else, I have found this book, like many others including yours, represents a projection, given some basic assumptions and patterns. There are many such, as you know better than I, and I don't expect that any of them will more than touch upon the actual course of events for the probabilities are imponderable.

That was pretty much the end of *Canticle* in our correspondence. Technological advance was a villain in Miller's work. Present at whatever level, it would become self-destructive. Efremov, as he often stated, felt that the basic danger lay in the "linear logic" and syllogistic approach of science, and the line between this and the technological villain, within the human vector, is very fine.

In the social area he felt that moral decay lay at the base of all declines, the failure of ethical standards to support the potential values of technology. The development of a pervasive "monoculture" kept bothering him; the problem of racial equivalence was an irritant. This was part of the cultural narrowness he found in the *Canticle*, necessary, I would think, to preserve a simple story line, but not a basic philosophical aspect of the main theme.

As a result of our correspondence, I had asked him a question upon his views on environment and its relationship to the spiritual differences of people. It turned out he felt strongly on this, and I received the following reply.

> About your environmental problem, you have asked me personally if I have that very sort of slant that there is a real spiritual difference between intelligent men of different races. On low levels it is non-significant (more or less low difference) and visa-versa, more high-more real difference and misunderstanding. Therefore all demagogy and cry about equality (especially in managing and government) and it is the biggest mistake of our time. So I think affairs will be worsened every year because of childish minds that cannot understand that simple law, which so clearly understood our fathers: rights inevitably suppose the responsibility (no responsibility—no rights). The quantity of unresponsible men increased very rapidly together with the appalling demands of rights. And this hook is very dangerous for every government which came on the path of false liberty. But I hope to discuss the very important and significant matter with you in person.

This didn't happen, which was unfortunate, for his letter was not too clear. Further, it didn't really answer my query, but raised some other points, clearly those of an elitist, at least a necessary pose for more than mere survival in the Soviet Union. Are then cycles of freedom coordinated with cycles of loose morality, irresponsibility and holocaust? Is egalitarianism fatal? One Russian certainly felt this to be the case, and the disastrous peak of the cycle late in the 1990s, which is what he foresaw, would be the culmination, the new Dark Ages, hopefully followed and setting the stage for a resurgence in the style of the *Canticle* cycles.

Or Progressively Up?

One comes up with a bleak outlook through Efremov's eyes. Yet in view of his dialectical concept of progress it is hard to see the future as unalterably bleak. Once he remarked to me that if we could get by the 1990s safely, all would be well. Then, however, there is the mystique and dangers of his peaks in the cycles. The fact is, of course, that while it is the fashion to talk of doom, day-to-day living requires some sort of confidence that doom is not real. Somehow or other, things will work out. The alternative is personal chaos. The latter was no part of Efremov's makeup. He knew, as well or better than most of us, that the worst could happen. He knew his heart limited his life span and alternately seemed to hope he would and would not see the onset of the "Dark Ages." Like most of us, he either ignored the possibility of this unfortunate time, tried to do something about it, or, for the most part, didn't really think it would come.

His major science fantasy writings and novels carry quite a different flavor, a confidence that the dialectical approach to knowledge will emerge and be successful. We saw some of this in his science and the concept is central in his most ambitious work Лезвие Бритвы or—*The Razor's Edge*. He notes in his explanatory introduction, as I have translated it, "The whole novel shows the special significance of knowledge of the psychological existence of man at present for the preparation of a scientific basis for education of people of a communist society." Three separate episodes, set in different places and with different characters each carry the same message.

The Razor's Edge (written without knowledge of Somerset Maugham's *The Razor's Edge*) is the thin line on which rests a delicate counterbalance of the arts and psychology and of science and technology. Psychology forms the spiritual link in the sense of Efremov's materialistic view of "spirit" as a human quality. Together, the balancing forces emerge in a dialectical spiral which, by inevitable progress, leads to an "ideal" society. It will, of course, be communistic. One or the other of the opposites, alone, can lead only to disaster. Efremov wrote to bring the message to the Russian people. The printing of 300,000 copies, as he remarked to me in an earlier letter, was far too low to meet the demands, and the black market flourished. No more were printed. Why, I can only guess; most likely, this was the number planned and a number not to be changed.

During the Stalin era, so Efremov said, all psychic and psychological studies and records were eliminated. Now, so we are told, they have become a tool of repression. Khruschev opened the doors once again to this form of enquiry and, during the 1960s and 1970s, there seemed to be a flow of strange "validated" psychic phenomena from the Soviet Union: sightless sight, mind impressions and images on photoplates and so on. But these are not Efremov's sense of psychic, psychology and psychiatry—his are manifestations of the materialistic human spirit. Spirit, rather, is one end of a materialistic continuum, a consequence of internal organization which exists in its own right. It is akin to "soul" (psyche) of Aristotle. It is not, and cannot be, independent of body, but on the contrary is the whole process of living. Expressions in the arts on the one side and in science on the other must be linked by this spirit or soul for upward progress to occur. This seems to me to be the essence of the message in *The Razor's Edge,* clothed as it is in the fanciful and the travails of the characters in the several episodes portrayed.

Many of Efremov's more strictly science fiction and science fantasies were based on a realized communism, a society stable and functioning. The vision is appealing. *Andromeda,* alternately rendered *The Nebula of Andromeda,* was translated into 35 languages, including English. It is a rather crude space novel, full of heroic adventures, intrigue and "good guys" and "bad guys" even on earth, where the story is anchored. *The Hour of the Bull,* a late novel, pits an earth-born communistic society, happy and content, against an anachronistic, capitalistic society developed from a much earlier earth society and "lost" for eons across the "null" zone on a far distant planet. It is an intricate, and for anyone less than fully at home in Russian, a difficult novel. The nice, patriotic analogue of east and west today is cast far away in space and time, accessible only by fanciful, time warp techniques of the far future (if at all). Efremov also employed the "space warp" drive in his *Cors Serpentis—The Heart of the Serpent.* Amusingly, this fanciful story somehow got identified in the *Bulletin of the Society of Vertebrate Paleontology* as a serious study of reptiles. Such has been communication. In most novels, Efremov stayed with clumsy, time consuming time travel and electromagnetic transmission, limited by the speed of light. His "Great Circle" used in several stories was an intergalactic web in which the nodes knew only that which had happened millennia earlier.

The ideal communistic societies were not without dialectical stress and strife—necessary, both to the philosophy and also to make a story. But no problems were too large to solve, even emergencies, by discussion and decisions of the appropriate Soviet in the framework of dialetical comprehensions.

At no time was I able to get from Efremov his own resolutions of his different points of view. Much like an evolutionary Jesuit, he seemed to package them under separate covers and did not try for synthesis. The closest he came was in *The Razor's Edge*, but here, for all the adventure, there is really none of his sense of imminent doom.

Writing, as I have noted earlier, was an escape for Efremov, escape into places where he could become the heroic figure in a romantic setting. He really did not want to go deeper. Early, it was pure adventure, geological expeditions and sea voyages. Soon, as early as 1945, it was stellar ships. Later, financial gain entered in and sometimes science was put aside for writing as his financial needs grew along with his illness and his disenchantment with the Paleontological Institute. Probably, the themes and development were not totally immune to this financial factor. I do know that he was most apprehensive about *The Razor's Edge* as a source of trouble, and was greatly relieved when it first appeared, uncensored, in segments in a periodical. It was not consciously made palatable. I feel that his "brave new world" was a hope, rather than a dialectical certainty. But too, it gave a mirror of hope to his readers, bogged deeply in the controlled bureaucracy. Away in space and time things could be said that were not possible in contemporary contexts. Science fiction writing during these decades was probably consistently the freest form of writing in the Soviet Union.

13

Books, Writings and Ideals

An Answer

From our first meeting, Efremov's wealth of knowledge and understanding of the world and its people intrigued me. He knew a great deal about the culture, the diverse peoples, history and politics of the United States, not just history book facts, but a sense of the problems of the different parts of the country and the critical aspects of both local and national politics. At this time, in 1959, the racial problems that surfaced at Little Rock had stirred wide interest in the Soviet Union, and the first question I was asked after a talk in Leningrad was, belligerently, what about Little Rock? Efremov somehow was aware of the deeper aspects of this matter. I had speculated that he had a short wave radio and that his knowledge of several languages might be an explanation of his broad knowledge and understanding. But this may or may not have been true, and his knowledge was not that to be gained from overseas broadcasts anyhow. What might have been apparent to me at the time of my first visit to his old apartment, in what appeared to be a done-over factory, dawned on me only much later. This apartment, like his later one near Moscow University, was lined with shelves of books. He read everything he could get his hands on in Russian, English, French, German, Italian and Spanish, and I expect some other languages as well. Both his breadth of knowledge and understanding, and the sometimes strange selective twists that his thinking took, were the product of this smorgasbord of reading superimposed upon his mixed social background, his travels during geological exploration through Asia, and his early romantic encounters with the sea.

The major themes in his non-scientific writing—the dominance of dialectical balance and sources of knowledge, ideal communism, the equation of man and intelligence, a goal of happiness, history, heroism and a pervasive, creative gentle eroticism—form the olio blended on the substrate of omnivorous and largely unselective reading.

Most of the books that he read could not be obtained in the Soviet Union and much of what he could get there he felt to be trash. He would pore over booklistings in magazines, ads and the books he received, cull out interesting titles and send "wanted" lists to his network of friends around the world. Although there was really no way he could repay the favors, many of us scoured the markets and sent along what we could find. I did, Mrs. Alfred Romer did, and so did Mrs. D. M. S. Watson of London and Richard Van Frank of Boston. Book dealers knew him well and helped where possible, but usually he could not pay them for their services and books. I am sure there were other persons in many countries who received and responded to his lists. In my case, when I was in the Soviet Union, Efremov did "repay" me by aiding in the purchase of gifts for my family, payment that was not at all necessary because the real value lay in his friendship. Most books were sent freely and without any thought of obligation.

The "wanted" lists were a hodgepodge and, in all, totalled hundreds of items. One that I received in 1971 is fairly typical.

> James Crazier, *Runt of Cygni*
> R. Valum, *Taurus Four*
> Jane Gaskell, *The Serpent*
> Michel Millet and Jean L'ange, *The School of Venus*
> Sax Rohmer, *Daughter of Fu Manchu*
> Ary C. Phillipes, *The Garden of Earthly Pleasures*
> Suzan Yorks, *Agency House, Malay*
> Allan W. Ross, *Bombay After Dark*
> Agnes Newton Keith, *Three Came Home*

Other lists contained a variety of such books as:

Lost Horizons
On the Beach
Darwin and the Naked Lady
Peyton Place
Man-Eaters
Forever Amber
Genghis Khan
Photographic Annual
Fantasy and science fiction magazines

and so on.

These, and many like them, were to be sent by mail. Air mail registered was the safest way. Some that I sent, such as *Mystery Magazine*, were intercepted because they were not licensed. They were merely taken out of the packet, and the rest of the books delivered. Books that could not be sent had to be carried in. This never posed me any problems for at no time were my bags opened.

One reason that some books would not get through was that the censors simply appropriated them, at least so I was told by Efremov. *Epoxy* was one of these. It was a moderately erotic French hardback "comic book" about the escapades and adventures of a member of a tribe of Amazons, females of heroic anatomy little concealed by scant cover. Efremov explained that he needed it as a basis for illustrations in his novels, being unable to obtain proper models in the Soviet Union. He had asked a British book dealer, Mr. Alan Myers, to try to get it for him from France. But, of course, he could not pay the book dealer for the book or for his service. So I was to send the needed funds to London, $10.00 if I recall rightly, and the dealer would buy the book and send it to me. Then I would carry it to Efremov when I next visited Moscow. I did, having been both pleased and amazed at this artistic and outlandish brand of comic book. Pornographic comics are a stock-in-trade in many countries, but this was rather an artistic, erotic adventure story. I would have thought it would not be allowed in, but Efremov assured me that the only problem was that the censors would steal it. It would not be rejected on technical grounds.

Many books, however, would have been officially rejected, either because they were not on the acceptable list or because they might raise some suspicions. Copies of my scientific monograph *Late Permian Vertebrates of the USA and USSR* did not get through on the first sending, as mentioned earlier. They were registered and so came back. I gather the USA-USSR did it. I sent them again and a more "liberal" censor must have been on duty, and they went through. *On the Beach* took two tries as well.

At the time that I was going to Russia to study, Efremov would send me "to bring" lists. These give some idea of what he thought would not pass the censors, for one or another of the above reasons. One list I received in 1961 included the following:

> The "Jap" romances, which I like very much
> R. Morley, *A Majority of One*
> J. Michener, *Sayonara* (Michener's works would not make it, I gather, so I took several. Efremov "ate them up.")
> V. Sneider, *Tea House of the August Moon*
> N. Beddiaeff, *Russian Idea*
> A. Burgess, *Small Woman*
> M. Pusie, *Top Secret Mission*
> I. Shaw, *The Young Lions*
> L. Poca, *History of Eroticism: de Erotica*
> Babarelle, *Adventure of Jodelle* (another erotic comic book)

Of the hundreds of titles he requested, I was only able to find a small percentage, for paperbacks come and go rapidly, with the old found mostly in the garrets of bookstores; take your pick—in the 1960s—at 10 cents a pound. Patterns are not at all clear in Efremov's lists. History, adventure with an erotic flavor, and sea romances stand out. There is a strong bias toward Asiatic stories. Concerning a package I had sent, Efremov wrote,

> Thank you ever so much for the books. You know I read every good and exciting book. So I accept with pleasure your gift. I have tried to read Faulkner [which I had sent, unsolicited, as an example of our literature] but I am not mature in dainty language nor a philologist and I find Faulkner flows through my mind in vain.

As for mysteries and detective stories, it seems to me that this sort of literature is coming to an end over the whole world with every year loss of readers. The historical novel with good erotic are now again arisen and have more success than before. Have you noted this peculiar phenomenon? I think it comes in the face of the apprehension and uncertainty in the current life conditions in the damned atomic age.

Gentle Eroticism

Science fiction tends to be short on character development and anything but the most superficial romanticisms. Of course, there are exceptions. The emotional parts of Efremov's stories are somewhat subdued and often rather awkward from our point of view. But his books carry a consistently erotic flavor. His longer stories blend sexual tensions into his full concepts of the meanings of life and man, intimately tied to the erotic qualities of music and art. This unusual combination in science fiction and fantasy is critical to the deeper framework of his studies of the future and distant places. They form a medium for his deeper ideas. His library book plate and other figures he enjoyed, illustrated in figures 31–33, are suggestive of his latent tastes.

In our conversations never was there a hint of preoccupation with the erotic. It could be that, in his writing, Efremov was catering to what he perceived as a popular demand, necessary to assure a wide audience for his work. I doubt it. Rather, reticence in conversations probably reflected his preference for "the old ways" in personal contacts, in line with the use of formal names. Perhaps the reputed Victorian or Puritanical Russian treatment of sexual matters, our stereotype, did enter in. I never saw anything to refute this, but then a casual observer likely would not.

Requests for *Forever Amber, Peyton Place, Epoxy* and so on prompted me to send one of our "vulgar books." It came about through a slight misadventure when a friend, Paul McGrew, and I had lunch in a Washington, DC, bar. Our well intentioned plans to see the city via Gray Line Bus Tour ended in a burlesque house. The show was predictably routine, jokes a hundred years old, water squirting in improper places, awkward strips, and bumps and grinds with appropriate drum beats. As always, during the intermission, candy, watches, and "girlie" picture books were hawked. I bought a book for 25 cents and gave it

Figure 31. The book plate Efremov used in the 1950s.

Figure 32. The fanciful image of a great, carnivorous dinosaur from the Cretaceous preserved in a photosensitive mineral slab. From *Stories—Shadows of the Past* as translated into English (1954)—a title in the Soviet Literature for Young People series. The illustration is by G. Petrov.

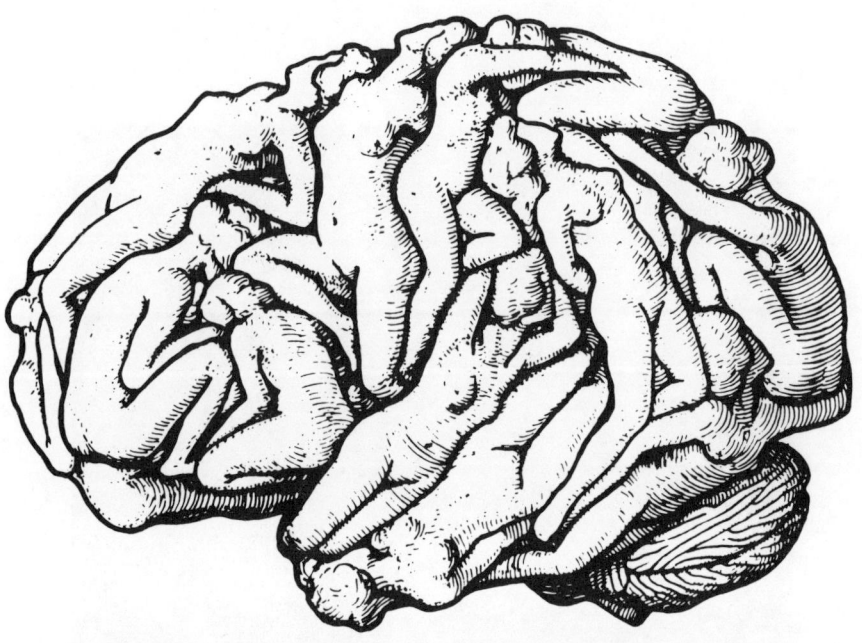

Figure 33. The "human brain" as portrayed by the use of nude females. A remarkable reproduction of the external anatomy of the brain representative of Bystrov's imaginative humor.

to Peter Chudinov, who was in Washington for a meeting, to take to Efremov. He took it back to Moscow. In due time, Efremov wrote to me and I could hear his strong Russian accent and stammer as he answered,

> Peter brought me the Frolics Book. What on earth is the matter with American men! Fat cows, leaning over to be milked. In Russia we prefer the broad hips!

He did have a healthy interest in the slightly vulgar, but his erotic sense was more sensitive and idealistic, coupled with his love of beauty, which found its finest expression in the perfect human body. His book plate, his love of ballet and gymnastics, and abhorrence of contact and violence in sports are all in this vein. His novels often raise erotic tensions, but just as quickly lower them in the face of social and moral discipline. This is nowhere clearer than in his story *Cors Serpentis*.

Far from earth, some 78 light years away:[4]

It was a wonderful world, but man, in his insatiable desire for knowledge had reached the very chasms of cosmic space, searching the solution to the riddle of the Universe. The space warp ship, the latest triumph of human genius, made it possible to answer the call of distant worlds.

"Yes, there's the other side," Kari Rami [a man] said aloud, unaware that he had spoken until Moot Ange's deep resonant voice singing an old song brought him to with a start.

The other side of love
Now rolling deep as the ocean's flood
Now narrow as a winding star
There's no escape, it's in your blood

Around the swimming pool of the starship, a sexual tension builds as divers, "their tawny skins gleaming with the glint of bronze that only a healthy outdoor existence can give, plunge in physical contact into the water." Tensions and momentary jealousies arise between the physically perfect males and females, build during waking periods, always giving place to inbuilt disciplines.

"Wouldn't it be wonderful, Kari, to find some secret passage on board?" The speaker was Taina, a tall, slender girl in a short tunic of shining green fabric that matched her eyes. Often she irritated the staid, level-headed Kari by her impulsiveness.

"Leading to some mysterious chambers where . . . ," Kari replied.

"Yes Kari, go on . . . ."

"That's as far as my imagination goes . . . . "

But Taina had got into the spirit of the thing, she pulled Kari after her into a dimly lit passage. An unfamiliar emotion seized him as he took the girl's hand.

"Let's go to the library," he said. "I still have two hours before my watch."

She followed him obediently. The air grew vibrant as the door opened with the tumultuous sounds of electromagnetic viono.

---

[4]The quotations are from a translation of *Cors Serpentis*, somewhat edited. In a few places the continuity has been slightly altered to give the basic flow without extensive attention to the non-essential.

"It's Moot Ange," she said, pressing Kari's fingers.

The music flowed in intervening harmonies . . . . Just then the door opened and Afra Devi, the doctor, slipped into the room to report an emergency . . . .

Two space ships from different parts of the galaxy—one with an oxygen-based life, the earth ship, and the other with a fluorine-based life—had made a close approach encounter, exchanging information, although no direct personal contact was possible because of the antagonistic life support media. Near the end of the meeting . . .

The stranger switched on terrestrial lighting and the earth men turned off the blue light. Two of the strangers, a man and a woman, threw off their dark red clothing and stood naked, hand in hand, before the earth men. The harmonius proportions (of their bodies) accorded fully with the earthly concept of beauty.

Their heads sat proudly on their long necks. The man had broad shoulders and the general physique of a worker and fighter, while the wide hips of the woman in no way jarred with the intelligent power that emanated from these inhabitants of the unknown planet.

The terrestrial light went out. At the commander's request, Tey Eron and Afra Devi stepped up hand-in-hand before the transparent partition. Their superb beauty caused a sharp gasp of admiration from their comrades. The strangers too seemed similarly affected. They looked briefly at one another in wonder and exchanged brief gestures.

"Now I have no doubt that they know what love is," said Taina. "True, beautiful human love—since their men and women are so beautiful and clever."

So much of Efremov is in these passages!

Ideal Dreams

Sometime in the far future, an ideal state of man, not the cataclysmic rises and falls of "Leibowitz," will be attained through the dialectical processes which are the structure of reality. Such is the deepest hope and the message that runs through many of Efremov's major, non-scientific later works. Man and intelligence are one and the same and, where one arises, the other must be. No force-field intelligences; no melding in an Omega Point of a matterless, Teilhardian spiritual skein; no monsters and no ruling robots.

Again, *Cors Serpentis:*

Only now, perhaps, did the astronauts fully realize that the driving force of all their searches, dreams and struggle was the good of Man. The most valuable thing in any civilization, on any star, in any island universe was Man, his reason, emotions, strength and beauty—his life!

Man's happiness, preservation and development was the main prospect of the future—now that the 'Heart of the Serpent' had been vanquished and there was no mad, ignorant, malicious waste of vital energy as there had been at lower stages of development.

Man was the only future force in the Universe that was capable of acting intelligently at overcoming the more formidable obstacles and advancing to a rationally organized world—the triumph of all-powerful life and the flowering of human personality.

Merging everything that he was, Efremov had this magnificent dream.

14

What of Dreams—Now?

Over a decade and a half has passed since I last spent time with Professor Efremov. He is gone, and so are Ernest and Wade. We still visit the red beds of Oklahoma and Texas, making small discoveries and fitting in the pieces of the ancient course of life on the once great delta near the Permian landmass.

A period of 15 or 20 years is such a short time in history, even in man's written history, that it can hardly be of any great significance. Still, if the studies of the ancient past can be trusted, it would seem that sometimes seeds planted during a very few years become catalysts for major events that follow, often chaotically non-predictable. The effects of some monstrous cataclysm that may have hastened the extinctions of the great reptiles of the past and paved the way for our own group, the mammals, seems to be one example. The "big bang" origin of the universe is another. But these are dramatic, and much more subtle events, recognizable only in retrospect, may also lie at the base of pervasive change. The early forays of our remote ancestors in search of food on the ground, or slight changes in the suspension of the upper jaws in bony fishes, fundamental to the great spread of modern day teleost fishes in the waters of the earth, are of this kind. One cannot presume to know that the decades of the 1970s and early 1980s have spawned such seeds, although there are those who feel this to be true, but so much has happened during this time as we plunge blindly at an accelerating pace into the future that the acceleration itself is at least a cause for concern about the fate of the old ways.

The course through our times seems to travel along two contrary pathways, the one marked by a burgeoning and overwhelming complexity of human affairs and the other an impelling roadway to comforting simplicity. The elated hopes and scientific dreamers of yesteryear, once public property, have now carried beyond common comprehension, parcelled out in necessarily simplistic abstractions by willing but handcuffed interpreters. Ideas, for the most part expressible only in mathematics, lose their intrinsic sense and mislead in articulate language. But the language and the world of science is, now more than ever before, mathematics girded by obscure acronyms. Perhaps a new literacy is emerging from the rise of computers. We seem to have made the circle back to the Pythagorean-Platonic world of "idos."

The dreams of the few who extrapolate from the equations into simpler dimensions, and their followers, seem to fight a losing battle, for now at least, against something mysteriously dubbed "nature" and a headlong, sometimes guided, withdrawal into myth and mysticism. Science, objectively searching for "reality," has been probing once "forbidden" areas, at least in our society, coming closer to the ultimate reduction of all phenomena to a common minutia of particles, ultimate atomism, energy and time, where forces strong and weak merge confusedly and are sensible only mathematically. Reductive analysis seems to flaunt holistic concepts both in biology and physics. Do we touch the unanswerable? Or is the very human source of the construct itself a bar to evaluating it? We unravel the mechanistic fabric of life and its origin, and life itself without a subjective sense seems nothing but a particular manifestation of a special set of molecules and catalysts. By treating, rearranging and putting together foreign fragments we alter old forms of being and create new ones. Yet, at the same time, in science we retreat from mechanisms and causes into a probabilistic or stochastic world of laws and chance. Is this the stuff of new dreams? Perhaps, but the puppeteer can hardly but confuse his audience.

A unity pervades the universe. The same matter-energy complex and the same laws apply throughout, and the lumps of matter and spots of energetic outburst are perturbations in a slowly subsiding burst. The "ultimate" origin of our universe in a "big bang" looms more and more probable. But what in the "instant" before, or what after? Or was there such an "instant?" Allegedly, by some, the "big bang" is the equivalent of the Biblical command, "Let there be light," as the struggle to bridge the gap between scientific and revealed knowledge finds it all to be clear in the Bible, the Koran or some other source of ancient interchange between God or gods and mortal intermediaries. Perhaps this conflict of the two is necessary for comfort.

None of this, of course, is new. The roots of our present scientific era date back to Copernicus and Galileo, followed by DesCartes, Newton and Darwin and recast anew by Einstein. And they go back to the Ionians through Muslim intermediaries. Each new surge of ideas has been met by movements to retreat into the mold of the past. Only recently, however, have moon landings, planetary explorations and radio and X-ray studies of the outer reaches of the universe, with its quasars and black holes, perhaps galactic cores, shown us so graphically how insignificant and possibly alone we may be on a temporary speck of dust in a seething maze of energy and near absolute zero "empty" space, all of which too must cease. Must we dream of other intelligences, even populate our skies with their vessels, and explore extragalactic future homelands to stay sane? Or shall we dream of intelligent robots, succeeding their imaginative but less intelligent creators and penetrating where carbon-based organisms cannot go?

Or shall we say, "Let's stop this, it's false; we live day-to-day and year-to-year and man is better for not knowing." Or again, "Let's remember, it is all filtered through our receptors and transmitters and created out of our own being, just a necessary construct to our existence and not necessarily real at all." The universe, thus seen, is our own creation, to die when man dies. Such strong disillusionment, such alienation, somehow seems to have intensified after the initial elation following the moon landings. The small window to space turned inward to the disturbingly isolated white and blue sphere of earth. Dreams began to dwindle. Grand space dreamers still exist—Ray Bradbury or A. C. Clarke among the fiction writers, Carl Sagan and John Ball among astronomers, Stephen Gould and Isaac Asimov among biologists. And to me, Professor Efremov, as well, was one of these. But his dream was covered by his own odd dialectical blanket and was totally anthropocentric in the sense of universal mission. All intelligence must be in man-form. Now, however, even some of the writers of science fiction and fantasy, who form the free, fanciful vanguard of the future, have tended to turn inward, following and even setting the earth-centered, humanity-centered trends. Beyond the few grand dreamers are the religious cults of a thousand sects, meditation, terrorism, scientific creationism, extremes of wealth and poverty, cultish environmentalism and the flowering of comforting myths of many sorts. Together they form a devastating mix with lines not clearly drawn, but mostly one way or another seeking explanations and reasons for being in semi- or pseudo-scientific sources.

Has all of this, which seems to have intensified during the last two decades and to be heading for a climax, any significance beyond today or tomorrow? Can we find in it any of those vague seeds which will flower either to roses or weeds? The obvious key to significance, which dates somewhat farther back, may lie in the added factor of man's newly acquired capacity to unleash immense amounts of energy in great, uncontrolled blasts. This casts its shadow soberingly over all phases of our life, both physical and social, and likely lies at the base of some of the restless searching for refuge in myths. A present danger to our "speck of dust" which did not exist before now looms as a power to destroy and becomes ever more a common property.

I would dearly love to sit with Professor Efremov, in his booklined little apartment on Gubkine Street in Moscow, with a "few drops" of cognac and talk of these things as we once did. Or, if not to talk, to write and get his reactions in his colorful, often obscure English. I can't, of course, but I might be permitted to imagine some of his comments, knowing all the time that they reflect my own sense of the words he might use.

——"Haven't you know yet," he might have begun, "not to make a drop in the sea a collision of earths? But I agree we head for troubles. Remember my chart? Your leaders have lost their ways, and like confused people, want to go back on old roads. Morals are at a steep decline among the young, too. We too, in Russia, go back to the old destruction of intellectuals and have no value of person or life. But maybe a new wind blows, maybe, but think back to Khruschev. Stalin destroyed the balance, just like your military-economic complex. I think our bad governments have no way to understanding the use of the damned atomic bomb. Not our countries, but maybe some crazy mass culture like Chinese.

——"I like your mixture of cults, terrorism, science, sociobiology and mystics, but don't you read my *Edge of the Razor*? It's all there, the linear mixup of undigested opposites with nothing of understanding of balance of the two sides. We must merge sciences and the arts and psychology. But I am horrified by your 'scientific creationism.' You wrote earlier, but I hardly believed it could be today, certainly not here. What on earth kind of balancing is this? What is the matter with you there, can't you kill them off dead like old Texas outlaws? What on earth is this sociobiology? You have sent me the book by Wilson. Has it passed now? He is a very wise man, but like you western scientists sees just one side. Maybe he is too much with ants. The twin studies you sent me from *Science,* how the pairs behave so much alike though long separated is puzzling—what about the others who don't behave alike? Are all these ways built in the genes? Maybe we have got Ardrey and Lysenko mixed together. You can't breed out 'cors serpentis.' It will go away as society not genes is so made over that the need of a person to win in combat is gone. But this is 'heady stuff' and you probably can't understand my badly put wanderings.

——"You keep talking about retreat to myths. This is very wise, but remember *Homo* is too young and embryonic in new consciousness to exist without them. Your western science and ours, too, is full of the greatest myth, that reality can be expressed and explained in mathematical equations. It's going the wrong way, we can pick out anything from the complex and prove it, but alone. The two sides of the equation are not equal explainers, but what I call 'linear tautologies.' It's dangerous, of course, without balance and synthesis. You say we are rushing headlong into the future and I agree, but on one-track rails without merging the parallel tracks of science and technology and arts and psychiatry. To the one end of the material scale, the spirit is missed. Tell me what you think.

——"Humanity and spirit you say show in kindness, charity and an effort to make all men equal. Now you retreat to the 'Great American Myth' like we French say *egalité*. This is a monstrous falseness. Look back through your 'sociobiology.' Ask your retreaters to nature, our modern Thoreaus, to look to their model. Where is the guide? A horrid danger—as I write you—is in the growing monoculture, the attainable *egalité* reduced to the level of the least. No! Spirit, soul if you want, is all life, not just man, and the task of evolution is to accomplish this end of materialistic and dialectical scale. Spirit is the key to the meetings. Being and awareness balance perfectly the other mechanistic end and then the goal of homeostasis is reached. Not at a low level. But not yet.

——"You drew back or came to ignore, quite right, your 'mathematical animals' in your *Morphological Integration*. Stochastics and mathematics only are helpful tools in getting to reality, but what of history? It never went a one-line path. With mathematics you make the universe what you want and the choice is in your bases and rules, mostly man made. We lose ends and aims; homeostasis is not a one-track process.

——"You are wrong, I am not one of your dreamers, but a 'turned in' earth dreamer. My stories, dreams if you want, are put in the future or the past, in outworld space, but only in fantasy. The ridiculous space warp must be, but just to make possible the return to earth of travellers with their lessons. Relativistic travellers can't come back like that. There would be no one to share their wisdom. Once, at first, I did write to escape my boring, but later mostly to tell my people things they needed to know and could not hear in our insane destruction by what you call communism. But you are right, man must have grand dreams, even if only of God and Heaven and now we seem to have lost them and even our formal religion is a misleading farce. I came down with the same worry in my *Edge of the Razor* but showed the way out. This was with some personal worry with me too, but nothing came from it. It will be a long way ahead, but like I have wrote you so many times, it is a dangerous trail and maybe only you of us will see to which way it goes. I think it to be soon, maybe in the 1990s?

——"But enough of this, you must be tired of my wandering. I agree things are changing very fast, too fast. But you are sad, not excited like I know you? Do you work too hard and maybe lose dreams too? Maybe, like I, you get old, too old. Maybe it is just today. But it is good to wonder and dream. Right now, it's a beautiful day out the window, but don't forget reality is only the moment. So let's stop and have just a few more drops and this time I say to hell with the heart and I join you . . . "

. . . Old Efraim

Index

A Canticle for Leibowitz 152–155
Africa 65
Anderson, Robert 35
Andromeda (The Nebula of Andromeda) 158
Arroyo formation 8–9, 15–19, 22, 37, 63–64

Barker, Maude 41–42, 48–49, 55, 57–60, 78
Barker, Wade ix, 41–49, 54–59, 64, 66, 78, 111, 173
Benjamin, Texas 18, 51, 53, 65–68, 75–78
Butterflies 3
Bystrov, A. P. 129–132, 168

Choza formation 9, 16, 18, 37, 51–52, 63–66, 73, 91
Chronofauna 64
Chudinov, Peter 99–102, 119, 130, 134, 168
Coffee Creek 17, 19, 21–22, 36
Cors Serpentis (Heart of the Serpent) 117, 158, 168–171
Cotylorhynchus 70–71
Covington, Ab 32, 34–35, 39–41
Cruthirds, Ernest ix, 17–36, 38–39, 41–42, 48, 55, 111, 173
Croneis, C. 5

Darwinian Centennial 144
Darwinism 132–133, 136
Davitaschvili, L. Sch. 124, 132, 135–143
Dialectics 123–129, 139–146
Dimetrodon 17, 40–41
Dobzhansky, Th. 124
Driver's Ranch 68, 75–77
DuBois, Ernest 22, 81

Efremov, Ivan A. ix–xiii, 3, 7, 10, 48, 79, 81, 86–94, 97–106, 111–135, 138, 142–176
　correspondence 85–91, 135–142
　cycles 148–152
　dreams 170–176
　Eroticism 165–170
Estemmenosuchus 86, 97, 99

Fergusen, Byrdie 34, 36
Flerov, Konstantine K. 95, 99–100, 103, 130
Fulda, Texas 18, 21

Gilliland, Texas 18, 42, 61–62, 72
Graham, Loren 123–124

Halsell's Ranch 56
Hotton, Nicholas, III 36, 43–44, 61, 66
Hour of the Bull xi, 158

Ignorant Ridge 18, 32–33, 37–38, 42–43, 52, 58, 67

Johnson, Ralph 32, 44–45, 47, 68, 71, 74

Kahn, Jack 68–69
Kahn Quarry 45, 68–70, 76
Konzhukova, E. D. 104
Krylov, B. 82–84, 94

Lake Kemp 17–18, 20, 25, 27
Lassen, Rick 54
Leningrad 116, 161
Lysenko 82, 123–124, 133, 138–139

MacFayden Ranch 66, 67

Michurin 138–139
Miller, Robert (Bob) 33, 44–45, 64, 89, 122
Miller, Paul 6, 17, 19, 40, 64
Miller, Walter M., Jr. 152–153, 155
Morphological Integration 89–91, 121, 178
Moscow
  arrival 1
  impressions 97–101

Neodarwinian theory xii, 143
Nitecki, Matt 69, 71, 103

Obruchev, P. 90, 95, 100, 103
Orlov, Yuri 2, 81, 83, 89–90, 95–96, 99–100, 102, 106, 115, 117
  correspondence 92–94

Paleontological Institute 81, 92, 95, 106, 159
Paleontological Museum 1, 8, 81, 85, 95, 97–106
Pangea 12–14
Patterson, Bryan 32, 64
Pease River 18, 66, 68
Permian ix, xi, xii, 1–2, 6–16, 38–39, 87, 121, 134, 173
  Soviet Union x–xii, 6, 8, 10, 65, 79, 88, 91–93, 99, 103
  Texas 2, 9, 17–18, 32, 40–41, 44, 56, 65, 68, 73–74, 79
Plate tectonics 11

Razor's Edge (Edge of the Razor) xi, 141, 143, 157–159, 177–178
Read, William (Bill) 22, 36
Reinhart, Roy 22, 36
Rigney, Harold 17, 19–20
Romer, Alfred S. 3, 6, 28–29, 40, 65, 81, 92, 100, 140
Romer, Mrs. Alfred S. (Ruth) 162
Roth, Robert 63, 66
Rozhdestvenski, A. K. 104
Russia 10
Russian language 80–82, 89–90

San Angelo formation 18, 52–53, 63, 65–68, 73–74, 79, 87
Seltin, Richard (Dick) 60, 69
Seymour, Texas 18, 20–21, 34–36, 41, 48, 78
Sharov, A. G. 104
Sharvar Tank 31–32
Shishkin, M. A. 104
Simpson, George G. 140, 144
Sternberg, Charles 37–39
Sternberg, George 27
Storms 57–62
Sukhanov, V. G. 101–104, 106
Sushkin, T. P. 115
Swanson, Vernon (Swanny) 60, 69
Synthetic theory xiii, 127, 136–137, 142–144

Taphonomy 88–89
Tatarinov, L. P. 100, 103
Terentjev, P. 146
Texas talk 22–23
The Dialectical Biologist 125–126
The Status of Evolutionary Thought in the West 135, 143
Trofimov, V. A. 100, 104
Truscott, Texas 18, 66, 68–75

USSR, Embassy 82–85
University of Chicago 71, 78, 81, 122
  geology and paleontology xii, 3–7, 44
  Russian course 81

Vale formation 9, 16, 18, 37, 63–64, 91
Van Frank, Richard 162
Vertebrate Paleozoology 90

Waggoner Ranch 17, 20, 29, 32, 34–35, 39, 41, 51, 55, 56, 61
Watson, Mrs. D. M. S. 162
White, Ted 37
Wichita Falls, Texas 21, 63